How to Be

Your Own Contractor

and Save Thousands on Your
New House or Renovation While
Keeping Your Day Job

With Companion CD-ROM

By Tanya R. Davis

How to Be Your Own Contractor & Save Thousands on Your
New House or Renovation While Keeping Your Day Job — With
Companion CD-ROM

Copyright © 2008 by Atlantic Publishing Group, Inc.
1405 SW 6th Ave. • Ocala, Florida 34471 • 800-814-1132 • 352-622-1875–Fax
Web site: www.atlantic-pub.com • E-mail: sales@atlantic-pub.com
SAN Number: 268-1250

ISBN-13: 978-1-60138-004-3   ISBN-10: 1-60138-004-6

Library of Congress Cataloging-in-Publication Data

Davis, Tanya R., 1962-
  How to be your own contractor and save thousands on your new house or
renovation : while keeping your day job, with companion CD-ROM / by
Tanya R. Davis.
     p. cm.
  Includes bibliographical references and index.
  ISBN-13: 978-1-60138-004-3 (alk. paper)
  ISBN-10: 1-60138-004-6 (alk. paper)
  1. House construction--Amateurs' manuals. 2. Contractors--Selection
and appointment. I. Title.

TH4816.D3347 2008
  690'.837--dc22

                              2008012672

COVER & INTERIOR LAYOUT DESIGN: Vickie Taylor • vtaylor@atlantic-pub.com

Printed in the United States

Printed on Recycled Paper

Dedication

For my three children, Heath, Amy, and Holly. You are the wind beneath my wings.

We recently lost our beloved pet "Bear," who was not only our best and dearest friend but also the "Vice President of Sunshine" here at Atlantic Publishing. He did not receive a salary but worked tirelessly 24 hours a day to please his parents. Bear was a rescue dog that turned around and showered myself, my wife Sherri, his grandparents  Jean, Bob and Nancy and every person and animal he met (maybe not rabbits) with friendship and love. He made a lot of people smile every day.

We wanted you to know that a portion of the profits of this book will be donated to The Humane Society of the United States. *–Douglas & Sherri Brown*

The human-animal bond is as old as human history. We cherish our animal companions for their unconditional affection and acceptance. We feel a thrill when we glimpse wild creatures in their natural habitat or in our own backyard.

Unfortunately, the human-animal bond has at times been weakened. Humans have exploited some animal species to the point of extinction.

The Humane Society of the United States makes a difference in the lives of animals here at home and worldwide. The HSUS is dedicated to creating a world where our relationship with animals is guided by compassion. We seek a truly humane society in which animals are respected for their intrinsic value, and where the human-animal bond is strong.

Want to help animals? We have plenty of suggestions. Adopt a pet from a local shelter, join The Humane Society and be a part of our work to help companion animals and wildlife. You will be funding our educational, legislative, investigative and outreach projects in the U.S. and across the globe.

Or perhaps you'd like to make a memorial donation in honor of a pet, friend or relative? You can through our Kindred Spirits program. And if you'd like to contribute in a more structured way, our Planned Giving Office has suggestions about estate planning, annuities, and even gifts of stock that avoid capital gains taxes.

Maybe you have land that you would like to preserve as a lasting habitat for wildlife. Our Wildlife Land Trust can help you. Perhaps the land you want to share is a backyard— that's enough. Our Urban Wildlife Sanctuary Program will show you how to create a habitat for your wild neighbors.

So you see, it's easy to help animals. And The HSUS is here to help.

THE HUMANE SOCIETY
OF THE UNITED STATES.

2100 L Street NW • Washington, DC 20037 • 202-452-1100

www.hsus.org

CONTENTS

Introduction

Welcome to the world of self-contracting. Whether you are building your first home from scratch, remodeling a home you already own, or investing in property to fix up and flip, this guide will be a great how-to manual to help you reach your goal. Building a home has become increasingly complex, but by developing a straightforward, organized way of handling each step of the process, you can easily create the structure you always dreamed of from beginning to end. You will be able to create a house that will be beautiful and functional, and bring you joy for years to come. The best part is that you will have the satisfaction of having created it yourself.

You may be reading this book because you have a dream. Whether your dream is to remodel your current home or build a new one, you want "something different." You may want more space, more light, or a better view. You may have attended scores of open houses, only to find that none of them quite suit you.

After realizing that there is no perfect house on the market, many people begin to consider tackling homebuilding themselves. This brings on many questions: Can I do it? Is it too confusing and mysterious? Is it too difficult? And most important, will it cost me more money than if I hired a contractor?

This book was written to help you answer those questions. It will help you decide whether the homebuilding process is right for you. It will help you turn your dream home into a reality: a custom home that you can be proud of.

Every owner-builder has three objectives. The first is to build his or her dream home, or transform the current one into the perfect home. The second goal is to save money, both by self-contracting and by making good decisions about supplies. The third is to save time; by streamlining the process, you will reduce time on the job as well as frustration.

This book attempts to address all three of those goals. It shows you how to begin and provides you with checklists to use for the major parts of the process. It helps you determine who carries which responsibilities, and how to make your site safer and more organized for your subcontractors (subs). It is not a step-by-step guide to, for example, lay flooring; there simply is not enough room to discuss every detail. Instead, it covers the contracting process and how to manage subs and materials — people and things — in an efficient way.

Whether your goal is to build one house or many, this book is a reference that is handy enough to carry along with you. It answers hundreds of questions to help you become a successful owner-builder.

YOUR OBJECTIVES

We both know that your objective in reading this manual was not just to sit down and read a good book. Instead, you are reading it to find ways to lower your costs, save time, and to eliminate as much frustration as you can from the project. Rather than tell you step-by-step how to build a foundation in tedious detail, this book tells you the six general steps to creating the foundation, and lists the possible mistakes you could make. It explains the most popular kinds of foundations. It tells how to site foundations on varying slopes.

In every chapter you will find ways to save. There is an entire chapter that is devoted to getting more for your money on the cost of materials. There are details about the bidding process: how to find contractors and ask for bids, how to select and eliminate possible contractors, and how to find ways to get contractors to do the job for less.

This manual should give you the knowledge you need to manage your own residential construction project. You will be able to work on the home yourself as much or as little as you want. And even if you never plan to lift a hammer, this book will guide you through the contracting process. By following the steps in each chapter, you can be organized and complete the process quickly.

It will help you avoid tasks that are best left to the professionals. It will also show you how to identify and avoid problems that other homeowners have encountered. It details the best floor plans, the way to choose a site, and the way to decide how to position the house on the site. It covers ways to create maximum enjoyment for your own family while helping you streamline the process, investing less time in it.

By building your own home, you are going to be intimately familiar with every detail of the house. You can arrange the floor plan to suit your family's lifestyle and activities. You can move the garage to the other side, or even to the back. You can choose the thickness of the outer walls and the kind of wood for your beams. You can use high-quality, durable products that will save you time and money in maintenance and repairs. This is a far cry from the average homeowner who approaches a builder asking him to build a "custom" home. These homeowners have little to do with the details, leaving them to the builder. They only select square footage, paint colors, carpeting, and perhaps some of the kitchen details. By comparison, when you are finished with your home, you will know every component. You will know why you used the materials you used, and you will have a house that is much more satisfactory to you.

Most of all, the book focuses on how you can save money on the construction of your own home — yet keep your day job. Many people believe that because they hold a 40-hour-a-week job, there is no way to oversee the construction process. This is simply not true. Plenty of people have learned to manage the project via the phone and e-mail, and spending weekends and evenings at the job site.

The ultimate success you have will depend on your own management skills, personal skills, and do-it-yourself skills. Residential construction projects normally involve 20 to 30 specialty tradesmen and each has his own specs, terminology, and expectations. In this book you will learn the main concepts for each trade so you can ask intelligent questions, gather bids, and select the right subcontractor with confidence.

THE FOCUS

Some construction manuals on the market today describe every detail of each aspect of building. Some focus on general building terms while devoting 30 percent of the book to getting financing. Some tell are about an author's one, specific experience building a home. Some are not worth the paper they were written on. But nowhere on the market was there a book like this — one that focuses on the two things you need to know the most: how to save money and how to juggle building or remodeling with your own job and family responsibilities.

I have not relied on my own experience as the only guideline for this book, although I have built and remodeled several homes myself. Instead, I have interviewed contractors all across the country to get tips and suggestions. Every contractor shared the name of at least one other contractor, who I then called and started asking questions all over again. I have wandered through construction sites and drilled everyone I know who has built and remodeled homes. I have combed the Internet and read every book I can find about building, the whole time trying to find ways that you can save money. As a result, there are tips for saving money in every section — real tips recommended by real people.

At the end of the book, you will find appendices full of helpful material, such as Web sites and magazines, that can help you learn how to be a builder. In addition to talking to builders, subs, and suppliers as I have done, I recommend that you read, read, read. The more information you have, the more confident you will feel about taking on the task. Relax, have fun, and enjoy your new home.

1

Are You Ready to Tackle the Custom Home Process?

Homeownership should not be out of reach and it should not be only for the rich. Owning not only "a" home, but "the" home — your dream house — is attainable for everyone. The secret is to build it yourself or to remodel your present home so it becomes the custom home you have always wanted.

There are many myths surrounding contracting. Whether they came from members of the construction industry or uninformed individuals, they continue to plague the do-it-yourself world, making most people think that contracting is a closed business. But building your own home is not only possible, it is smart. By building a home you save money — up to 35 percent compared to purchasing a previously built home — plus you are able to create the house you always dreamed of. And by building it yourself, you can be absolutely sure that the structure you create emphasizes the things you want it to. Plus, building your own house is loads of fun.

CAN YOU DESIGN YOUR OWN HOME?

Designing your own home. This brings up one of the biggest objections people have to building: the fear of creating a design. But you have been in and out of homes all your life — by now you have developed a firm sense of what you like in a home and, of course, what you dislike. So who better than you to select how your rooms will flow, what types of rooms you will have, how much storage each one has, and what the view will be from the windows? Building your own home gives you the freedom to create exactly what you want.

Even if you suspect that the only taste you have is in your mouth, there are ways that you can have exactly the home you want and build it yourself. For example, you can hire an architect to create the design you imagine. That way, you will not create a home so specific to you that it cannot be sold later. An architect can also help you identify potential problems in the dream house design you are carrying around in your head. Also, if you cannot form a clear picture of what you want, an architect can begin a design and modify it to suit your taste. He or she can help you figure out what type of home is best for you.

MYTHS OF BUILDING YOUR OWN HOME

Myth #1: High Materials Costs

One of the most common myths about building a house is that contractors can get better prices on materials because they buy in bulk, so a private individual loses a lot of money by buying materials. This is simply untrue. Most contractors do have special pricing with the dealers they use; however, these are not the only suppliers available to you. Contractors are also middlemen, so they may mark up the materials rather than charging you the price they get from the suppliers. It may be necessary to shop around, but good deals can be found. For example, if a contractor sends you to his lumberyard and you get a price of $18,000, talk to other builders and get estimates from their suppliers. You may save a few thousand dollars simply by being a savvy shopper.

Myth #2: You Have to Get a License

Another myth about owner-builders is that they are doing something illegal. Of course you must own the land and follow the local laws by getting all the right permits, having inspections, and eventually occupying the house, but there is nothing that restricts you from building (except in certain restricted subdivisions). Owner-builders do not need a specific license in most states, and there are no special restrictions unless you are attempting some of the more technical trades like electrical work or heating and air conditioning. You should investigate getting licensed as a builder, though. One lady found out that all she had to do was take a one-day class and she would have the credentials the lenders wanted to see. Having a license may help you get better prices and more funding.

Myth #3: Subcontractors Hate Owner-Builders

This myth is just silly; subcontractors who agree to work for you clearly do not have a problem working for private individuals. And if one happens not to show up, it is not because of you or your demands. It is because that subcontractor is unreliable. You can bet he did not show up for some other employer before you. This book attempts to de-mystify the hiring process as much as possible so that you can create a good relationship with the subcontractors you hire. It is vitally important that you interview the subs and check their references, so you will be assured that your workers will show up when they are supposed to.

Myth #4: You Cannot Do It

It is true that you may not know everything you need to know to build a house, but this problem can be easily overcome. There is no reason you cannot contract your home. In reality, contracting is less about the actual construction and more about management skills. It is always possible to hire outside help, and many people will be willing to give you advice when you need it. Your job will mostly be to organize and document the entire process. You (or your spouse) will have to spend a great deal of time at the job site, and you must pay close attention to detail. There will be something to do every day of the project, and most nights you will go to bed thinking about what you need to do the next day.

But ultimately, if you are wondering whether you can manage your own project, the answer is: of course you can. Thirty-five percent of the homes in this country are custom homes, and those are not all built by professional builders. The following are some things that will make your project run more smoothly.

WHAT IT TAKES TO DO-IT-YOURSELF (DIY)

To build or remodel a home, especially if you are doing it while keeping your day job, you must be willing to learn. If you have never participated in the construction industry, or if you have knowledge of only one area, you will need to learn new terminology, the specifics of building, and much more. Anyone who wants to undertake such a project will need to read, read, read. In addition to books like this one and the ones listed in the bibliography, read magazines about building. Some of the better ones are:

- ✗ *Home*

- ✗ *Home and Architectural Trends*

- ✗ *Taunton's Fine Home Building* (or anything by Taunton)

- ✗ *This Old House* (they also have a Web site with excellent articles)

Besides the fun reading and great design ideas, there are other things that owner-builders must read and examine. You will have to study the local codes, and be sure that you comply with them. You will also need to have at least some understanding of blueprints and their significance in the construction process.

The good news is that aside from reading and learning about the industry, you need only average organizational skills to act as your own contractor. It is great to have skills in any of the trades, but it is not a requirement.

YOUR RESPONSIBILITIES

The main job you will have as a contractor will be planning. This involves scheduling workers to be on the site to perform certain tasks, while at the same time coordinating the supplies and equipment they may need. If you have hired someone to operate a backhoe, for example, make sure you have rented and received the backhoe before he or she arrives to work. The last thing you want to do is pay people to stand around and wait for equipment. The better organized you are, the more money you will save.

To keep the task from seeming overwhelming, break it down into a series of steps, and proceed with each step as if it is a single project. If you feel that your organizational skills are lacking, read Organizing for Dummies (Eileen Roth and Elizabeth Miles). You could also look online at **www.napo.net**, a Web site that helps people get organized.

The second responsibility of a contractor is to estimate costs. You will have to remain within your budget if you want the loan to cover the cost of construction. Otherwise there will be cost overruns, which will force you to find money in other places like your assets or savings account. It is a good idea to keep some additional funds set aside for emergencies or any problems that occur. You do not want a cash-flow problem to keep you from finishing your house on time.

One important responsibility a contractor has is to find good subs. If you gather a roomful of do-it-yourself builders, you will hear tons of horror stories about subcontractors. Complaints like, "They do not show up," "They are always late," and "I have to stand over them." You will quickly realize that getting good subs is going to be the key to getting your job done on time. If you are careful to hire the right people, they will do their job without close supervision, and you can spend less time on site. Chapter 6 will cover how to do this.

Last, you will be expected to organize the job. As the contractor, you are the boss. No one will do his or her part without a directive from you. The more organized you are and the better you stick to your budget, the smoother the job will go. The workers will follow your lead, and they will complete their jobs on time without wasting resources.

YOUR ATTITUDE

The nature of a contractor's job demands having certain expectations of the workers and suppliers he or she works with. However, your attitude will play a huge role in determining how the contractors you hire behave. Your attitude will also affect the way that you feel about the job — and about yourself. By communicating well with everyone involved, you can make sure that your expectations are understood. You will also troubleshoot many problems before they arise. No one wants to work for a rude boss, so projecting a positive image will boost morale, and compel your workers to be effective and efficient.

CAN I SAVE MONEY?

There are many fallacies about homebuilding, and you are sure to have heard at least one or two. For example, you may have heard that owner-builders do not truly save any money compared with contractor-built homes. This is absolutely not true. Take a survey of 10 people who have built their own homes or have done their own extensive remodeling and you will find that every one of them saved money. In fact, many will say that they saved around 50 percent.

You can assuredly save money by building your own home. This is understandable because when a contractor builds for you, he is doing it to make money. His or her work always includes "salary" even though the structure of the contract will not call it that. Most builders mark up their jobs by at least 10 percent, so you can look at 10 percent of the cost as the absolute minimum you will save.

Ten percent of a $300,000 home is $30,000 — quite a savings, especially when you consider the amount you would pay on that in interest over the life of your loan. But if you use the tips in this book as a guide, you could save more. If you saved 35 percent on that same $300,000 home by building it yourself, you would have $105,000 in savings.

Now let us look at remodeling. Your total budget will probably be smaller when remodeling. But the savings can still be astronomical. Say you own a home worth $298,000. You decide to remodel, converting your garage into another bedroom and while you are at it you add on a 560-square-foot addition, which enlarges and opens up your kitchen. Acting as your own contractor, you turn your three-bedroom home into a four-bedroom with a new kitchen addition. You spend about $66,000 on the project. You saved between $6,000 and $7,000 by overseeing the project yourself.

But wait — you decide to refinance the home after the addition is complete, so you order an appraisal. The house that in your mind was worth $364,000 (that is the $298,000 plus the $66,000 addition) appraises at $413,000. The value of your house has increased by $115,000 — almost double the expense of the remodeling project.

Remodeling an older home instead of building a new house can be advantageous if you find that you are not ready to build your dream home. You can remodel the one you own, or look around for a home in the neighborhood you want to live in that needs to be fixed up. These are often advertised as "diamonds in the rough" or some other terminology that indicates a less-than-perfect appearance. Sometimes a low-interest loan exists on the property that can be assumed, making an older home especially attractive. If you are in a hurry to move, a remodeling project can be completed months before a complete new build would be. Plus, some areas offer tax incentives for remodeling.

COMMON PROBLEMS

There are a few problems that crop up again and again with custom home building. Some of these are due to "other people" — people who tell you that you cannot do what you are doing. Some of the problems are due to the fact that you are trying to juggle a home life, a job, and a large construction project. Some are simply due to lack of experience. These are a few of the more common ones:

Lack of a license — Do you need a license? Normally special licensing is not required if you are building a home for yourself, but check with your local building inspection department to be sure. Licenses are required for specialty trades.

No flexibility — If you are building a home from the ground up, you will need to spend at least part of every day on the job site. This may mean that you will have to set aside the time to be there. If your job and family life do not allow you to spend time at the job site, it is vital to determine alternate ways to handle problems and supervise your employees. For example, if a subcontractor phones with a question at 1 p.m., make sure someone can get there within the hour to look at the problem.

No time — If you find that things are getting too stressful and you have no time for your family, you might consider paying a builder for assistance. The builder can help you by attending the site while you are at work, lending you subs, and may even help you get a construction loan. Builders will often do this if they are not busy, because being present on your project might secure him another buyer.

Treating the project as if money is no object — Everyone has limitations.

To stay within them, it is vital that you set the budget before breaking ground, and then stick to it. The best approach is to visit the lenders first, before you have many details in your dream book about the house you want to build. You can make faster, more reasonable decisions by knowing your financial constraints.

THE ADVANTAGES

There are many advantages to building your own home. Most people who have done it mentioned that they enjoyed having total control over the quality of construction and the materials. Many contractors have a certain method they like to use or a certain way of doing things; they do not necessarily change their system to accommodate the customer. By building yourself, you can truly "have it your way."

The savings have already been addressed — most people save an average of 15 to 30 percent. This creates an instant home equity on the property. Plus, it will whittle away some monthly mortgage costs. Because of the do-it-yourself savings, your payment will be lower than if you had purchased a ready-built home.

You get to choose the plans. Builders often have a collection of "standard" plans, which may meet your needs. But they may not have a collection or you may simply want to customize the floor plan. A contractor will accommodate those changes — at a cost. If you do some research, which will be covered in Chapter 5, you will easily be able to find plans with a style that is exactly right for you.

Another advantage to building your own home is that you get to select the building materials that are used. As you research the construction process, you will learn about types of foundation, subflooring, and plumbing. You will learn that although there are "optimal" sizes of furnaces recommended by manufacturers, there are "better" sizes to use. Finding out about these factors causes most people to want to choose for themselves. By selecting your own materials, you can control the quality of the house, and emphasize the elements that are most important to you.

The best thing that comes from building your own home is the sense of satisfaction you will get when it is complete. Forming an idea of what you want your home to look like and then working hard to make that idea a reality will be one of your biggest accomplishments. The completion of the project — the culmination of all your hard work — will be the proudest moment of your life.

CREATE A WORKBOOK

To get started on your project, use a workbook. This can be a three-ring binder, either with the plastic pocket sleeves or without. Using the pocket sleeves gives a little more flexibility if you decide to move things around. If you are completely opposed to the binder, use an expanding folder with 13 sections. Also, set up a file drawer in your home office; the paperwork generated will be enough for two entire drawers.

In the accordion file or notebook, label these categories:

- Dream Project
- Design Plans
- Contracts
- Communication
- Financial Information
- Invoices
- Lot
- Materials
- Receipts
- Permits
- Subs
- Warranties

Under the "dream project section," go ahead and start saving magazine articles, paint-chip colors, and other tidbits you have gathered as you have thought about building your own home. Even if you are just beginning the journey, it will be an advantage later to have the ideas on hand. Having these images all in one place will allow you to mix up designs, and see how they work together. You will want to make sure that your colors and shapes work well together, and you will want to have everything together, allowing you to make informed decisions.

TRUE FINANCIAL ABILITY

After setting up a filing system, the first thing any owner-contractor must do is determine the budget for the project. Nothing can be done until you are certain of how much money you can spend on the house. This will largely be decided by the bank, but it would be best to sit down with your CPA to assess your finances first.

To get a loan, many people simply visit lenders, and they choose the one who offers the best package. The loan amount you can get is often different from the monthly payment you are actually comfortable with. To plan for an amount that feels right to you, define the factors that will have an impact on your home. Some of these are:

* How much cash you have on hand

* How capital gains may affect you

* Other assets

* Anticipated expenses, like upcoming college tuition

* Your long-term investment plan

* How long you plan to own the home

* Expected appreciation on the property

✕ The tax implications of your loan package (interest and points)

After you have considered these, it is time to sit down with a loan officer. In Chapter 4, we will cover in detail how to apply for loans and what types of loans are appropriate for owner-builders. A loan officer can help you determine the loan amount that you can afford; you can also use the online calculators at **mortgage-calc.com** to find out the amount of money you might qualify for.

SET A BUDGET

After the loan amount has been determined, set a budget for the construction process. This should not be 100 percent of the loan. It seems that no matter how careful you are, something always causes building projects to go over budget. It is best to set the budget at 90 percent, or even 80 percent, of your total budget. Setting aside additional funds will prevent hang-ups that could destroy your project. Running out of money could force you to sell assets or even to delay the project indefinitely.

DECISIONS, DECISIONS

The first really big decision you have to make is where you are going to build your house. If you are looking for land, you will have to research and explore all the options. Finding just the right land is not something you can do sitting at home in front of the computer. You will have to get out and hunt for it. You will have to contend with bad directions, mislabeled streets, properties that are not at all what you imagined, and much more. Consider it an adventure — just sit back and enjoy the ride. You will know when you find just the right spot.

CHOOSING A LOT: WHAT TO LOOK FOR

There is a difference between buying a lot and buying land. A lot, or a finished lot, is already prepared for a home to be built on it. It has some or all of the utilities, and the ones that are not present are accessible.

Land, on the other hand, is not ready for building. Raw land might be harder to finance — or the financing could be more expensive — because it requires extra work. If your heart is set on buying raw land, consider looking for private financing or ask the seller to carry the loan. Allow more time for the building process — raw land will require special permits, roads, utilities, and possibly drilling a well.

Do you select the lot first, or the style of home first? This is sort of a chicken-and-egg question. For now, evaluate the lot as if you have already determined that the house you are going to build will fit on it.

Money-Saving Tip:

If you want to save money, purchase a "finished lot" instead of raw land. Lots can be purchased with less money down, they are easier to finance, and you will spend less time prepping the property for construction.

Lay of the Land

When looking at lots and land, there are so many factors that the choices may be dizzying. Many of the choices are also extremely personal, and they will be determined by your lifestyle. The lot you deem "perfect" might be one your best friend would never consider.

A lot may be called "buildable," but what does that mean? Even if the city, county, or state in which the land is located has deemed it buildable, will it be economically realistic to do so? You will often find that people refer to a site as buildable, and they may even use that word in the advertising copy, but there may be good reasons not to build on it. To determine whether a site is buildable for you, make sure it meets every single requirement at every level of government. This is where it pays to dig. One experienced home builder told me that he had learned to go to government offices and ask repeated, non-stop questions — then go back the next day and ask some more. He learned the hard way that if you do not ask exactly the right question, you might not get the answer that you need.

Realistically, though, it is not up to government office workers to solve your problems. You are responsible for your own purchase, design, and solutions.

When evaluating a lot, walk every inch of it. Study the lay of the land, where drainpipes have been placed, and look for soft spots. If there are trees, many of them may have to be removed for construction. If there are rocks, you might have to excavate or blast them. Following are some other things to consider.

Slope

If your family spends a lot of time in the yard, you may want to avoid a lot with a steep slope. If the lot you are considering is sloped, you need to study the way it drains. Does the water drain toward the house site? Will it be difficult to mow? Will you have a hard time driving up the driveway in the winter? Does water collect on the property? Also consider the implications of a lot that is too flat. If the slope is less than 2 percent, then it may have drainage problems.

It is best to find land that is fairly level or that slopes gradually upward from the road. Of course, if you particularly like a location but the slope is not exactly what you want or need, it is possible to grade it and make it what you want. Factor in that cost with the total purchase price of the lot. In addition, if you are removing a lot of dirt you will have to pay to have it hauled away.

Privacy

Most people enjoy some level of privacy in their yard. Even if you select a home site on a busy street, you will want rear privacy and, depending on your personal lifestyle, possibly on the sides as well. Corner lots may feel less private, and they also have a shape that will complicate the placement of a home. Also, make sure the lot is big enough to leave a comfortable amount of yard space between your house and your neighbor's lot. Trees or fences can help create a sense of privacy, but you will need enough room in the yard to put them in.

City Versus Country

Are you more comfortable in the city or the country? The old saying about the three most important factors (location, location, and location) is not far from the truth; you are going to have to live there. Land outside the city will have less local taxes; land inside the city might have better access to roadways and highways. County land will be quiet and peaceful, but city land will be close to nightlife, restaurants, and shopping centers. Think about your lifestyle and what location will most improve your quality of life.

Another consideration when determining the location of your house is how much time you are willing to spend on maintenance. A couple of acres can cost a great deal in terms of money and time when you are mowing and landscaping. For some, this extra time is worth it; others would not dream of spending the entire weekend on land maintenance.

Subdivisions

One of the biggest advantages to choosing a lot in a subdivision is that resale is usually fast and profitable. In addition to that, home values will be somewhat stable because the homes are governed by covenants. Before purchasing in a subdivision, get a copy of the covenants. Be sure that they allow owner-builders and that they protect against the building of much lower-priced homes than the one you are planning. Do the covenants state when you can or must build? Do you have control over the time frame, or do you have to build within a certain time after purchase?

Also, pay close attention to homeowners' policies and subdivision rules. Many subdivisions have rules governing the appearance of your property and any additional work that you might wish to do to it. Of course, subdivisions will offer a more social atmosphere, and if you hate yard work or some of their other required maintenance, you can find a neighborhood teen to do it for you.

Style of Home

You probably have a favorite style of home, or a type of home that you always dreamed of owning. Upon examining your dream house file, it may have pictures of many similar homes — so many that you realize this is your "type." When looking at land, consider this type of home (the style you plan to build) and compare it with the homes that are in view of the proposed site. If you are building a log home, you do not want to purchase a lot in a subdivision full of Italianate structures. Likewise, if you are building an Italianate structure, you do not want to put it in view of a subdivision full of manufactured homes. While there is nothing wrong with having a house that stands out, you do not want to feel like your home is out of place.

Consider Resale

Many people build a home with the intention of staying in it forever. But others know that they will get bored, be transferred, or that the family size will change, and the dream home will have to be sold. For that reason, consider resale possibilities when purchasing your land. The most difficult land to resell will be:

⚒ Located in a flood plain

⚒ In or near a declining neighborhood

⚒ Near any of the following:

- Airports
- Railroad tracks
- Landfills
- Electric lines
- Power easements
- Radio towers
- Major roadways

Not only do these factors affect the resale value of the home, but they can also influence your life in that home. Late night sounds from trains and airplanes may drive you crazy, and who wants to deal with a flooded basement? You also do not want to have the most expensive home located in a low or mid-range neighborhood. It can make your home more difficult for a future buyer to finance.

Easements, Encroachments, and Other Criteria

The different government authorities have criteria that have to be met, both to lessen the impact on the environment and to satisfy the residents of the community. When you find a piece of land that you like, check with the land assessor associated with the county to determine its legal status. He or she can show you the legal description of the lot, and can tell you whether there are any easements.

Zoning

There are several designations of zoning, and each has a different meaning and certain limitations. The zoning tells you:

- The minimum lot size you can use
- What kind of structure you can build
- How many units can exist on the lot
- What the property can be used for

Zoning information is available at the local government office. Go to city hall if the property is in the city, or go to the county offices if the property is in the county. Most lots for building houses are zoned R-1 (residential with one unit). A higher number means that more units can be built on the property. If you are looking at agricultural land, you might see a designation like RA-3. Rather than referring to the number of units, this zoning means that the residential agricultural property must be at least three acres in size. So, if you purchased five acres that was zoned RA-3, you could not subdivide later and sell off part of the property for building; each lot has to be at least three acres.

Zoning can quickly change. The local government is not required to check with you, obtain your approval, or even notify you about zoning changes ahead of time. The best way to check on zoning is to visit with the local government office to see if there are zoning meetings scheduled for your area. While there, find out about zoning of the surrounding properties and if there have been recent changes. Also, ask if there are any special tax assessments you should know about. Special tax assessments are charges that the government can place against your real estate for public projects. These include water or sewer lines, paving, parking structures, streetlights, and other assessments that property owners may need to fund.

When thinking about buying a lot in a subdivision, there may be more restrictions on building in the covenants than there would be in a freestanding lot. These restrictions can be obtained through the homeowners' association or through a title company. Examine them to find out setback limits, style limitations, height limits, and design review guidelines before putting an offer for the lot in writing.

Buildability

Now it is time to determine whether the site is truly buildable. That means whether the soil will perc, whether the land might flood, whether a septic system will be allowed, and so on. "Perc" means a percolation test, which is a soil test to see how fast water will drain through the soil. Perc tests are performed by septic specialists to see whether a septic system might be allowed. The county then approves the septic design.

If there are wetlands, creeks, or lakes on or even near the property, there may be certain environmental studies required. If the site involves a hillside, there will be a geotechnical analysis to determine whether it is a seismic hazard or landslide area.

Key Items for Buildability Verification

- Legal description
- Zoning designation
- Lot size versus setback
- Encumbrances, encroachments, right-of-ways
- Availability of sewer or an approved septic design
- Availability of electricity (also find out the requirements)
- Availability of natural gas (also find out the hook-up requirements)
- Availability of police and fire protection
- Driveway accessibility (does it require an easement?)
- Storm drainage requirements for the local area
- TV and cable service requirements
- Geotechnical analysis (is one needed?)
- Seismic area review (is it needed?)

- Covenants, conditions, and restrictions (if any)

Items Affecting Cost

- Extensive water, sewer, or electrical hookup
- Fire marshal's requirements (i.e., fire truck turnaround)
- Access roads
- Storm drains or other runoff control
- Wetlands or other environmental impact
- Removal of hazardous wastes (if present)

BUDGETING FOR LAND PURCHASE

How much should you spend on land? This is difficult to determine early in the game. A good guideline is 25 percent or less of your total budget.

Money-Saving Tip: How to save on the land purchase

When purchasing land, it is of paramount importance to know what the property is worth. This is part of the research you complete before the offer. As soon as you can, find out what the property's market value is by finding the prices of the most recently sold land in the same area. If the lot is located in a subdivision, look at the last properties sold there. How were they like yours? How were they different? What was the price? By comparing the desirability of your lot with those that have recently been sold, you can get a good idea what the market value should be.

Checklist for the Lot You Are Considering

- Does it allow for a basement, either by sloping or good drainage?

�֎ Is it on the sewer already? If so, is the lowest drainpipe placement going to be higher than the sewer line?

✖ Is it in a desirable location with a safe atmosphere?

✖ Is it in a low-traffic area?

✖ Will the shape be suitable once a house is built on it?

✖ Is it in a flood plain or full of subterranean rocks?

✖ Is the base soil or shale? Is it solid?

✖ How attractive are the surrounding homes?

✖ What is the view like?

✖ Is the entire lot usable?

✖ How is the area zoned? Is that suitable for you?

✖ Are underground utilities available (phone, electric, cable, and television)?

✖ Is the lot private?

✖ Will most of the trees and shrubs remain once construction is under way?

✖ Is it accessible to major highways?

✖ Is it convenient to shopping, schools, parks, etc.?

✖ How much are taxes per year? Is it inside or outside city limits?

✖ Is it located in an area of growth or decline?

One way to make the property work for you financially is to purchase it first and pay it off. Using the land as collateral, you then borrow money to build the house. This is called leveraging the property. It is the best way to ensure that you will get a construction loan. It is not the only way to get a loan — but it is one idea.

Things to Consider When Designing Your Home

1. Style: formal or casual?

2. How many stories?   _____

3. How many bedrooms?   _____

   ✘ Every bedroom should have two means of escape in case of fire

   ✘ Every bedroom needs a closet and easy access to a bathroom

   ✘ No walking through bedrooms to get to other rooms

   ✘ Every bedroom needs room on one wall for a king- or queen-size bed

4. Great Room   ___yes ___no

5. Living Room   ___yes ___no

6. Media Room   ___yes ___no

7. Mud Room   ___yes ___no

8. How many bathrooms?   ___

        Where?   _____

9. Where will the laundry room go?

10. Where will the master bedroom go?

11. Regular-Height Ceilings____ or Tall Ceilings ___

12. Eat-in Kitchen____ Formal Dining Room ____ Both_____

- A kitchen needs a clear walkway of 32" at each door and 42" for the work aisles. At least 60" of wall for cabinets is required, with cabinets being 12" deep and at least 30" high.
- Work centers — sinks and prep centers must not be separated by tall cabinets or the refrigerator. The entire "work triangle" (see Chapter 12) should be 26' or less. Each leg of the triangle should be between 4' and 9'.
- For an eight-person dining room, allow at least 10' by 12'.

13. Where will the storage areas be located?

14. How many garage spaces? _____

15. Hallways should be minimum length to save money, but they need a minimum width of 36".

CHAPTER 1 CHECKLIST

- Create a workbook with the proper categories
- Examine your financial ability including current assets, debts, future obligations, and income
- Set a budget for the land and house
- Select a lot considering:
    - Lay of the land including slope and drainage

- City versus Country
- Subdivision versus acreage
- Style of home you desire
- Resale potential
- Zoning and restrictions

🛠 Verify buildability

🛠 Complete a lot checklist

🛠 Complete a design consideration checklist

2

Plans

Most likely, the reason you decided to build your own home is because you wanted to design it yourself. You may have a markedly clear idea of the style and design you want. If not, there are several ways to go about getting house plans. Even if you are going to hire an architect to design it for you, I would strongly suggest that you study plans and consider them so that you accomplish two things:

1. Familiarize yourself with floor plans and how they translate into "real" rooms.

2. Determine the exact plan that will work for you and your family.

By now your "dream house" file contains a list of the requirements for your house. If it does not, go ahead and make the list before spending a great deal of time looking at plans. Your list should include what types of rooms you want, approximate sizes, and the number of rooms. That way, when you find a plan that you think you like, you can double check it against the "must-have" list.

PLACES TO FIND HOUSE PLANS

You can find house plans almost anywhere. They are in building, renovation, and home-interior magazines. They are in most libraries and bookstores. They can readily be found on the Internet. There are whole books and CD-ROMs of plans that you can purchase, either online or at the local builders supply store.

Sometimes plans can be ordered in sets; other times they are offered individually. If you are wondering why there are "sets," you will need one for yourself, one for your lender, and one for your contractor or subs. When you choose to put the project up for bid, each bidder will want a set of the plans to look at. Most people order about six sets.

Money-Saving Tip:

You can get copies made of the set of plans for about $10 each. So, if ordering extra sets is expensive, save money by having them copied at the local office-supply store, unless they are copyrighted.

The advantage of ordering one of the thousands of pre-drawn plans you can find is that you can stick to it without hiring an architect or designer to make changes. If you do have to make a change, hire an experienced draftsman, designer, or architect. Requesting changes to a pre-drawn plan will cost much less than having someone design the entire house.

Draw Your Own

Purchasing stock plans is the least expensive way to design your home, but you can draw your own plans if you prefer. The easiest way to begin drawing your plans is to purchase simple graph paper. Use the scale of 1/4 inch to 1 linear foot. Draw the plans the best you can, and then show it to your builder, architect, or designer. He or she can take your estimates and point out where you need to add or subtract items to reduce waste and save costs. Professionals will also be able to point out parts of your design that may be impractical or inefficient.

Another way to create your own plans is to purchase a kit. This kit will contain:

✖ Scaled grid sheets

✖ Appropriately sized furniture

✖ Parts, like windows, doors, and walls

✖ Landscaping items like trees

The kit may be either two-dimensional or, in some cases ,three-dimensional. It may include plumbing fixtures, TVs, stereos, and more. The kit can make it easier for even a professional to create the plans, so do not be ashamed or afraid to utilize it. The advantage of using a kit is that it makes it much easier to visualize the project as it will be when it is completed. Seeing a multi-dimensional view of your home will allow you to get a more accurate idea of what the home will look like, and you may come up with new, better ideas.

Software

Computer Assisted Design (CAD) engineering and graphics packages are available to everyone who owns a computer. The software will guide you through the process of designing your home. You can rotate the drawings on the screen to see all the angles of the room, and you can get scaled drawings that show the proportions of the elements in your home. Drawings can be saved and printed. The designs are surprisingly accurate, and they are so professional that you can forgo the cost of having a professional complete them. If you choose to use CAD, start out by taking the time to learn how to use the program. It is absolutely necessary to know how to enter the correct dimensions or you may ruin the flow of your design. Tutorials and help files are available for a reason — use them.

ARCHITECT VERSUS DESIGNER

Most municipalities require that plans meet certain Building Codes and standards. For example, in New York you must have the design certified if it is over 1,500 square feet; additions and remodeling over $10,000 fall under this requirement, too. This is done by a design professional — either an architect or a designer. But which one should you use?

Architects have more formal training. They study at least six years in school, spend a couple of years at internships, and then have to obtain architectural licenses. Architects also work with the Building Codes day in and day out, so they can save you a great deal of time and money.

An architect will be familiar with the special building considerations for your particular area. Ask to see his or her portfolio, and be sure that you use an AIA certified architect. Architect fees are structured at an hourly rate, a lump sum, or a percentage of the construction cost. Expect an architect to charge from 5 to 15 percent of the construction cost.

Designers can draw plans and also help you determine what materials are needed. Some homeowners like to draw the plans themselves and then hire a designer. Another option is to purchase stock home plans (a cheaper alternative than using an architect) and then employ a designer. If money is an issue, it is possible to save by hiring a home designer who is not a licensed architect. Besides design, they can help you choose the most durable materials for the least cost. Find a designer in the local newspaper or a home design magazine.

Architects and Remodeling

You and your subcontractors can handle remodeling projects that only address aesthetic changes, such as appliances, flooring, and countertops. However, anything that changes the blueprint, like changing the shape of a room or adding rooms onto the house, requires an architect.

Some states prohibit unlicensed individuals from designing entire houses, so make sure you check your state guidelines before you begin work. It is possible that unlicensed persons can complete smaller jobs, such as a remodel or small addition to your home.

BUILDING GREEN

Green building is the practice of reducing the impact of building on the environment. Key steps include specifying green-building materials from local sources whenever possible — generating at least some on-site energy, optimizing systems like heating and cooling, and recycling. Low-impact building materials may include low VOC-emitting materials and paint. VOCs, or Volatile Organic Compounds, are the cause of about 3,000 diagnosed cases of cancer each year. Using local materials eliminates emissions created by transportation, and it also supports your local community. In some cases, like heating and cooling, you will save money over time by going green because you will have much lower energy costs.

Many people who are intent on green building salvage materials from old houses — doors, windows, mantels, and hardware. This prevents old materials from ending up in a landfill. Using bamboo or cork oak will ensure that the wood you use is rapidly replenished. It is amazing how fast bamboo grows and it is a beautiful, strong wood.

Energy loads are controlled by using daylight, passive solar heat and thermal mass storage, and hot water recycling. These activities will help you save on your energy bill and they will also reduce your environmental impact.

Building green utilizes many interesting materials. One is straw bales, which are used for walls. They are covered with stucco and have an R-value of R-30 — nearly three times that of traditional wood stud wall systems. Higher R-values mean that the wall retains more thermal energy and is better insulated.

Green building maximizes some of the following:

- Efficient resources by using recycled content, materials from sustainably managed sources, or materials that are natural, plentiful, or easily renewable.
- Materials that are found locally or regionally.
- Items that have been salvaged from disposal were renovated, repaired, restored, or improved through remanufacturing.
- Reusable or recyclable materials, enclosed in recycled content or recyclable packages.
- Materials that are longer lasting as compared to conventional products.
- Improved indoor air quality, through low or nontoxic materials, which emit fewer carcinogens or irritants and minimal emissions of VOCs. Air is also improved via products and systems that are moisture resistant to resist biological contaminant growth.
- Materials and components that require simple, non-toxic, low-VOC cleaning methods and products that enhance the air quality or identify indoor air pollutants round out the green air system.
- Energy efficiency by using materials components and systems to reduce energy consumption.
- Passive solar design or other building shapes and orientations that utilize natural lighting.
- High-efficiency lighting systems.
- Properly sized and energy-efficient heating and cooling system.

- Minimize the electricity loads from lighting equipment and appliances.

- Alternative energy sources where appropriate.

Water conservation is another element of green building. Green building is important to water conservation because it minimizes waste water by using low-flush toilets, low-flow showerheads, etc. Often, green builders will use recirculating systems for centralized hot water distribution. It is possible to design for dual plumbing that will use recycled water for toilet flushing, or a gray water system that recovers rainwater for site irrigation. Builders may also use micro-irrigation to supply water in non-green areas and use self-closing nozzles on hoses.

One big objection to building green is the question of affordability. Keep in ming that these days, it is possible to afford much more than ever before. In addition, it is important to keep in mind that a cost now, such as solar panels, results in large amounts of savings over time. For people interested in environmental issues, there are ways to conserve both while building and for when the home is completed at almost every financial level.

Green Building Room by Room

Bathrooms

The United States uses 300 billion gallons of water every day. That is enough to fill Olympic-sized swimming pools placed end-to-end all the way around the earth. If water conservation is high priority in your area, consider installing low-flow toilets, showers, and faucets. Install a hot water recirculation pump with a timer. These recirculation pumps are convenient because you do not have to wait for the water to get hot; it is already hot in the pipes. Also, because the water gets heated based on a timer, energy is not wasted by reheating water that is not needed.

Kitchen

About one-fourth of the nation's largest water treatment facilities are in serious violation of pollution standards at any time. More than seven million Americans become ill each year due to contaminated tap water. If this is a concern to you, consider installing a distiller. These actually cost less than going to the store and buying bottled water and they only use about three KWH of electricity per gallon. Distillers remove contaminants from the water you drink.

Electricity

Electricity uses a lot of our natural resources. Most of it is derived from coal and other sources that have a large carbon footprint. To reduce your carbon footprint, try using "green" power or solar energy.

Local power utility companies that buy solar or wind power and provide it to you provide green power. The cost is slightly higher, but some areas offer incentives or tax breaks to offset the extra costs.

Incorporating passive solar design into a new construction is easy. Passive solar uses the sun to heat and cool your home. It can work on its own or be used in tandem with more traditional power sources. If you want to try passive solar energy, plan to face windows to the south to collect heat.

Solar systems can be on-grid (connected to the power utility) or off-grid. If you are building in an extremely remote area, consider off-grid systems.

The Air

Indoor air pollution is one of the top five environmental risks. It causes 14 times more deaths than outdoor air pollution. To reduce indoor air pollution in your new home, use low-VOC paint, seal duct work tightly, and insulate effectively. Visit **www.epa.gov** for more information about clean indoor air.

Green Floors

Earth-friendly flooring may be referred to as green, natural, or non-toxic. Many consumers are now demanding sustainable flooring because it will improve the environment and reduce toxicity levels in their homes. Earth-friendly flooring includes wood, bamboo, cork, natural carpeting, concrete, and natural linoleum.

Concrete — Concrete is not always considered a flooring "option" but it is inexpensive and can be painted, scored, or stained. Concrete is becoming more popular as a flooring choice. It can be cold to the touch, though, and it presents a hard surface to stand on if you are on your feet a lot at home.

Carpeting — Green carpeting refers to natural materials: coconut husks, sisal, sea grass, wool, and jute. These carpets are low in VOCs, are stain-resistant, and are made from recycled materials. Be sure the backing that has been used is also environmentally friendly.

Linoleum — Linoleum is non-toxic and biodegradable. It is made from natural materials like linseed, rosin, or jute. Linoleum can last up to 40 years and is a great alternative to vinyl, which may contain harmful pollutants.

Wood — Not all wood is earth-friendly. If wood is not forested properly, it may harm the environment. To purchase hardwood flooring that is "green," look for products that have been certified by the Forest Stewardship council (FSC). FSC-certified wood is available in all popular species, and some of the more exotic woods like teak. The FSC works to make sure that environmental pollution, habitat destruction, and wildlife displacement is minimal. If you choose hardwood for your floors, be sure to use low-VOC stains and finishes to reduce toxicity and the release of harmful chemicals.

Bamboo — This is an extremely renewable resource and a popular choice for flooring. It grows fast and easily with little fertilizer, pesticides, or care. Bamboo makes a beautiful, durable floor that is water-resistant. When you shop for bamboo, be sure your supplier knows its source. Formaldehyde may have been used as a binder — he will know for sure.

Cork — Natural cork is made from the bark of the cork tree. It is resistant to mold and microorganisms and it feels great under your feet. Plus, cork is great for reducing sound. Cork costs about the same as hardwood flooring and can be used for floors, walls, and underlayment.

Outdoors

To continue the green building concept on the outside of the home, use steel framing instead of wood. Steel products today contain 60 percent recycled content, compared to the 40-year-old trees that would be required to frame a 2,000-square-foot home. Fencing and concrete products made from recycled materials are readily available. Urban compost and mulch may be made in your hometown or nearby, and can be delivered when you are ready for landscaping. Salvaged brick and stone make use of existing resources without straining the supply of virgin resources. Also consider recycling your construction waste. Forty percent of the waste that goes into landfills is from construction; more than half of this is recyclable.

Green Remodeling

If you are interested in cutting the heating bill and at the same time reducing your carbon footprint, consider making some changes during your remodeling project. About two-thirds of your energy bill goes into heat; half of that is wasted. To offset this waste, consider insulating the basement, attic, and garage. Insulate the water heater if it is older, and wrap your exposed hot water pipes with insulation. Or consider replacing the water heater and other appliances with ENERGY STAR® equipment to reduce costs. Become extremely green by using under floor heating, solar water heating, and supplementing your regular heating system with a wood or pellet stove.

If you are interested in green construction, visit the U.S. Green Building Council (**www.usgbc.org**) or **www.doityourself.com**, which contains an entire collection of topic-specific green building articles. Learn about ENERGY STAR® ratings at **www.energystar.gov**. You can find green supplies at **www.greenbuildingsupply.com**.

PLACING YOUR HOUSE ON THE LOT

Siting your home is one of the most important parts of building. The way your house sits on the land will determine how sunlight enters your rooms, what sort of view you have when you look out the windows, and whether you have privacy or feel as if you are in a fishbowl. It is important to site the house in such a way that you will feel comfortable inside it.

Prior to deciding how to locate the house on your lot or land, be sure to get a site plan from the surveyor. This plan will show you the dimensions, easements, and setback requirements. By utlizing the site plan, you can place the house properly. If you do not, you might encroach on someone else's property or, worse yet, on a public easement.

Study the property's drainage pattern. Look for potential problems, like rocks or springs that might be located under the soil. These must be avoided, which may mean that you have to set the house differently than you originally planned. Also, take care to look at the way the property will drain after grading a building. If you change the drainage pattern to flow onto someone else's property, you can be held liable for damages. After you have located and graded the property, be sure it does not drain onto the neighbors' property. You also want to be sure that the water on your property will drain into an unneeded space. For example, gardens, driveways, and patios could be damaged by too much water.

To lay out the corner stakes, first determine which direction the house will face. Most houses look best when seated parallel to the street, but it is perfectly acceptable to orient it some other way — especially if you are concerned about sunlight for solar heating. Be sure that you honor the setback limits in the front, side, and back of the house. Measure the front line first, and then locate the back corners. Square the stakes by measuring the diagonals of the square — they should be the same. Then, to be safe, go back and measure all the setbacks again to be sure you are in compliance.

CONTENTS OF HOUSE PLANS

House plans actually consist of several parts other than the floor plan.

A Plat (Plot Plan)

You, the designer, or a surveyor will create the plat. It is a map of your lot with the house drawn in. Its purpose is twofold. First, if your house will not fit on the scaled drawing of the lot, it will not fit on the real lot. Second, the plot plan is required to get the building permit. To create the plat, first draw the lot, then sketch in all the zoning setbacks and restrictions. Next, draw in the house.

Spec Sheets

The spec sheet is a list of the specifications for the materials to be used. It will include the type and grade of wood, block, roofing, concrete, siding, brick, insulation, etc. The easiest way to create the spec sheet is to use an existing one, and add to it as needed.

Floor Plan

The floor plan shows the entire outside dimensions, location of windows and doors, and plumbing fixtures. The electrical, heating, and cooling fixtures are not included in the floor plans — they will be drawn out on separate sheets. There should be one plan for each level of the house.

Foundation Plan

The foundation plan simply shows the house and the location of load-bearing components like the footers, basement slab, piers, reinforcing rods, etc.

Detail Sheet

This diagram gives the specifications for footings, framing, and a layout of the kitchen cabinets.

Elevations

The elevation sheets show the outside of the house from all sides. There will be a front, a rear, and two side views. It shows the windows and doors (location as well as size) and indicates what kind of exterior materials will be used.

Special Plans

Special plans may include mechanical, electrical, and plumbing layouts, or these may be done by the subs. Additionally, there may be other detailed layouts like trim, tile, windows and doors, and the roof.

DESIGN CONSIDERATIONS

The lender might require you to build the house exactly as the plan was presented. Changes that are seemingly small can have a ripple effect — for example, adding a couple of feet in an upstairs bedroom will cause changes on the floors below — which can end up being extremely expensive. The best way to save money is to keep from making any changes once the plans are in place. Consider design changes in the early stages. Take your time and try to examine all the possibilities. There will always be items that you realize you want to change, but these changes should only be superficial. Changing the structure of the house after construction has begun is unrealistic. Typically, you will want to fine-tune details such as kitchen cabinets or which way the doors swing.

There are many more details to take into consideration than the book can reveal here. Only the biggest factors that many DIY builders have encountered are addressed.

Traffic Patterns

An acceptable traffic pattern pushes movement to one side rather than circulating people directly through the center of a room. If you have a floor plan that does this, consider whether simply moving a door toward a corner will solve the problem.

Storage

Some home designs do not have enough storage — even in designs regularly used by popular builders. Add bigger closets in the bedrooms than you think you will need and consider bumping out the garage by a couple of extra feet. Plan for storage in the attic or basement, and build a pantry into the kitchen.

Placement of Windows and Doors

Windows can let in unwanted heat, invite burglars, and turn your house into a fishbowl for the neighbors. Exterior doors give the first impression of the home; interior doors may block passageways or prevent large furniture from going in or out. Consider every placement carefully and measure doorways. Interior doors should swing into rooms rather than the hall.

Convenience

Think about your daily activities. How accommodating is the new floor plan for the things you do? Pretend that you are walking and living throughout the new plan. How does it feel? How convenient will it be to do laundry or carry in the groceries?

Building for the Future

Families go through stages, and as a result each family outgrows their home about every five years. If you can plan the house accordingly, you might save yourself from moving. If you have not yet started a family, consider building a three or four bedroom home. If you have several teens who will be off to college in a few years, try building a style of home that allows you to close off a portion.

Potential Problems

Just as beauty is in the eye of the beholder, whether a building issue is a problem is up to you. You are building your house. If you put enough thought into it, you can make it exactly what you want. However, when it comes to plans there can be hidden problems. Here are the biggest ones:

Deciding to build a certain house that you have seen in a magazine before finding a lot — Searching for a lot to fit plans that you already have can end up being very expensive. Instead, consider finding a couple of potential lots first, then finding plans that will fit them.

Choosing stock plans that do not match your climate — Stock plans that are purchased from a magazine or Web site do not take into account your climate or the nature of your lot. For this reason, you may want to hire a professional to make modifications, or at least look it over.

Wasted space — Long hallways are the biggest cause of wasted space. Once you have the basic plans laid out, consider the way you will walk through the entire space. Eliminate unnecessary wasted areas. Remember — you are paying by the square foot.

Doors that connect awkwardly — Like a bathroom that connects to a kitchen or a bath that can only be reached by guests walking through a bedroom. It is also awkward when the back door can be seen directly in front of the front door, or when the front door seems to lead to the stairs instead of the living area.

Master bedroom on the front of the house — If the house is on a street with any traffic, the master bedroom will seem noisy.

Building the biggest house in the neighborhood — This can make your house harder to sell.

BASEMENTS: HOW IMPORTANT ARE THEY?

Whether to put in a basement is an important consideration, and it is something you should give much thought. Part of the decision will be based on the area in which you live. A good, dry basement can be a bonus, but a wet basement will make you wish you never had one. A basement that allows enough water to enter the home can eventually cause the foundation to be destroyed.

If you want a basement, it is best to avoid flat lots. A sloped lot will allow you to create good drainage and avoid potential problems with dampness or mold. Sloped lots also allow for walkout basements, which have both natural light and ventilation possibilities.

Because basements can be problematic, consider whether you need one. If the water table is high or the land does not drain well, flooding can be an annual event. A damp basement may require a dehumidifier, professional waterproofing, or constant clean-up and maintenance work.

On the positive side, a basement is fairly inexpensive to build. It gives extra storage space and a place to put the heating system. It can also be great for a home office, workshop, or getaway den for teenagers.

DESIGNING TO SAVE MONEY

At the design stage, there are a few things you can do to cut costs. If after you have the basic plan you realize there will be cost overruns, consider some of the following.

Use a floor plan that is as near a rectangular shape as you can get it. This is the most efficient plan. It allows for a simple layout and a simple roofline.

Use a smaller footprint. Bumping out bay windows and breakfast areas several feet breaks up the straight lines of the house and adds architectural interest. They give you more sure footage without costing you anything for an enlarged footprint. The foundation will be smaller, so the cost is lower.

Consider building the house closer to the street. Driveways are expensive. You can save thousands on concrete just by moving the house on the site. Be sure to allow for the proper setbacks.

Consider whether you really need a basement. Basements can add 5 to 7 percent to the cost of a home.

Consider putting off some of the extras. By only rough finishing the basement or upstairs, you can save money now and finish it as needed.

Lower the ceilings. If you have chosen 9-, 10-, or 12-foot ceilings, you can save money on drywall, paint, trim, and heating costs by lowering the ceilings.

Consider combining rooms. Each interior wall costs money. If you can reduce the need for separation, you can save money and give the double room an airy feel.

Choose factory built homes, modular homes, or "panelized" construction. Panelized is a type of building that is somewhere between modular and stick built. Wall panels are pre-constructed with the windows, doors, siding, and insulation. They are shipped on two trucks, unloaded, and assembled. These homes are constructed in only a few days. When finalized, onlookers cannot see that it was a panelized house. Kit homes also offer advantages, even though they are not pre-constructed. If you purchase a kit home, you will know the exact price of the materials.

SEVEN STEPS TO THE PERFECT HOME DESIGN

1. Make a list of all your wants and needs.

2. Select your favorite three home designs as the final contenders so you have options for fitting on the lot.

3. Include the possibility of future expansions to your home, even if they are as minor as finishing a basement or attic.

4. Make sure you have room to move around even after you have put in your furnishings.

5. Kitchen traffic should flow well from the sink to the stove to the refrigerator and have enough counter space to work easily.

6. Consider putting the laundry room on the same floor as your bedrooms. It makes sense to not have to carry loads of laundry up and down the stairs.

7. If you have special hobbies like sewing, painting, scrapbooking, or playing music, consider incorporating your own playroom into the plan. It can be custom designed to fit your hobby, and the rest of your family may appreciate the clutter or noise being contained. One artist said her dream house would have a separate room for each type of craft.

CHAPTER 2 CHECKLIST

- Look at available house plans
- Consider drawing your own plans
- Consider having an architect or designer draw or modify plans
- Determine which of the above approaches you will use
- Consider to what degree you want to use "green" materials and techniques, and include that in your plans
- Determine the placement of the house on the lot
- Review your plans to make sure they include:
    - A plat plan
    - Spec sheets
    - Floor plan
    - Foundation plan
    - Detail sheets
    - Elevations
    - Mechanical, electrical and plumbing layouts (possibly by others)
- Check the floor plan for:
    - Traffic patterns (door locations)
    - Window locations

- Storage areas and closets

- Convenience for daily activities

⚒ Determine whether to have a basement

⚒ Consider the money-saving design ideas above

⚒ Review the seven steps to the perfect house design

3

Pre-Construction Activities

There are many activities that must be done before beginning construction. You have already done some: you started your notebook, began reading about construction, and created your house plans. Now it is time to move on to the pre-construction activities like estimating costs and gathering tools. At this point, the to-do list may seem endless. This is a process that must be accomplished one step at a time.

It is also a process that should not be rushed. Most contractors recommend that first-time builders spend at least as much time on planning as they do on building. That means if you are going to build your house in six months, you should plan for six months first.

Before beginning these activities, walk through your floor plan one more time. Make sure that the plans are exactly the way you want them. Making changes on paper is easy — but changes made to the windows after installation will cost you money, time, and labor.

THE STEPS OF CONSTRUCTION

This chapter covers the first major step — pre-construction. The other steps occur in the following approximate order:

- Excavation
- Foundation
- Framing
- Roofing/Gutters
- Plumbing
- HVAC
- Electrical
- Masonry
- Siding
- Insulation
- Drywall
- Trim
- Flooring
- Landscaping
- Painting, Staining, Wall Covering
- Cabinetry and Counters

SCHEDULE AND FLOW CHARTS

A written schedule with deadlines will encourage most people to get the project done on time. According to the National Association of Home Builders, the average owner-builder takes nine and a half months to complete a home. But with an organized, written plan, a reasonable goal could be six months. The best way to start is by writing out a timeline of how things will proceed.

THE MANY STEPS INVOLVED IN CONSTRUCTION

Construction involves many processes, and each one is dependent on one or more of the others. Activities may take more or less time than you expect; there may be scheduling delays and weather delays. Therefore, the entire plan has to be somewhat flexible.

It is also important to get the building permit, insurance, and other required preliminaries before creating your schedule. There is nothing more frustrating than delaying construction because there is no permit yet.

On the other hand, construction loans have pre-defined longevities. Some are four months, while others may be six months. So, your timeline will be influenced by the loan longevity. Know your loan terms before creating your timeline.

HOW TO FORM A CONSTRUCTION SCHEDULE

To begin creating a timeline, it is important that you lay it out carefully. Ask subs how much time they need for the job and how much notice they need when you are ready for them. Ask suppliers how much lead time they need to get your materials. Include those details on the timeline. The worst feeling in the world is holding up your own construction because you forgot to order something crucial.

After you think through your schedule and write it out, have your site supervisor look it over. If you are not hiring a supervisor, consider asking a sub or a supplier to review your written schedule. Either will usually be happy to offer suggestions for running the project more smoothly.

To create a construction timeline, it is easiest to use a computer program like Excel. The most important part of the timeline is to be sure that every single item is listed, and that the list is in the correct order.

When you are finished, take a step back and look at it. Does the flow make sense? If it does, go to each sub and get a firm commitment in writing. This should include a start date, an estimated duration for the project, and a finish date. Make adjustments to your timeline if necessary.

HOW TO FORM A CONSTRUCTION FLOW CHART

A construction flow chart is exactly like the timeline except it is a bar graph or picture instead of words. In a way, the flow chart is easier to "read" because you simply glance at the colored blocks.

FLOW CHART FOR ALLEN ADDICTION											
Phase	Name	Week 1	2	3	4	5	6	7	8	9	10
		Date 11/10	17 Nov	24 Nov	1 Dec	8 Dec	15 Dec	22 Dec	4 Jan	11 Jan	14 Jan
Site Setup	Self	■									
Deconstruction	Self, one crew	■									
Site Earthwork	AA Excavating		■	■							
Foundation	AA Excavating			■	■						
Concrete	DeLong				■	■					
Frame	Custom Framing					■	■	■			
Windows/ Doors	Self								■		
Stucco	Alberson			■							
Mechanical	Superior			■	■						
Plumbing	McDaniels				■						
Electrical	None yet					■	■				
Insulation	Carter						■				
Drywall	Belvins							■			
Cabinetry	Zibbers									■	
Finish Carpentry	Self								■	■	
Paint	Self								■		■

COST ESTIMATION

Estimating costs is one of the most time-consuming tasks you will perform. It is also one of the most important if you are constrained by a construction loan or spending limit. Spending time on cost estimation, and regularly spending time studying expenditures, will prevent cost overruns.

One important source of finding out costs is your subcontractors. Although you can get estimates from suppliers, the subs will have a better idea of what is involved.

The construction plans you purchased may have come with a materials list. If you have to pay extra for it, do not bother; these lists are often not reliable. If you do use it, treat it as a guide and run your own estimates.

When figuring your estimates, use the checklist in the back of this chapter as a guide. Take the numbers from the list and enter each onto a purchase order form for your various suppliers. Remember to control costs from the beginning. You will not save 10 to 35 percent of the cost if you do not control the spending.

There are a few things in particular to consider when creating estimates.

Excavation is charged by the hour. Larger equipment costs more to rent, but can perform the work in a short time. Clearing a half acre of land may take from one to four hours; digging the foundation might take around three hours.

Concrete estimates are figured on 4" slabs. Eighty-one square feet of slab requires one cubic yard of concrete. Footings are twice as wide as the wall, and there will be a local code requirement that determines the total dimensions. After checking the code, figure up the cubic yards of concrete needed, and include three extra feet of depth for every pier hole plus at least 10 percent for waste. If the depth is 12" and the width is 8", you will need 0.025 cubic yards of concrete for every lineal foot. These figures are for smooth finish work, not driveways or sidewalks.

Roofing from scratch is harder to calculate than other types of roofing. The plans should help you to calculate the rafter length and the roof area. The pitch of the roof is described as inches per foot of distance. In other words, a 10" slope rises 10" for every foot of horizontal distance. Once you have figured the square footage of the roof, add 1-1/2 square feet for every foot of eaves, ridge, or valley. Divide the total by 100 to get the number of squares. This is how roofing shingles are sold — in squares, which is how many shingles will cover 100 square feet of roof. Roofing felt goes underneath the shingles and is available in 500 square foot rolls. Divide the square footage of the roof by 500, and then add 20 percent for waste to determine how many rolls you need.

If you are going to use subcontractors to do the work, always allow them to give a bid to complete your estimate. It takes a lot of time to calculate your costs, and they may do it faster and more accurately. Always use the bid price over your own estimate, making sure you agree on the square footage of the area to be covered.

Surveys are required twice by the lender. The first one is a vacant land survey and the second is a mortgage survey. Even if you do not have a lender, it is a good idea to have the mortgage survey so that you have a record of the site, setback requirements, and where the house is located in relation to the property lines. This can prevent many problems later on. I remember one owner-builder who placed his house 15 feet over the setback. Because he did not have a mortgage, no one caught the mistake until the house was under contract, and the deal fell through due to the problem.

Water tap-in fees can be obtained from the municipality. If there is going to be a well, get a bid from a drilling firm. Be sure that their bid includes a maximum price and a per-linear-foot price.

CHAPTER 3 CHECKLIST

- ✖ Walk through the floor plan one last time and make any desired changes now

- ✖ Determine the construction timeline and create the construction flow chart

- ✖ Develop a materials list

- ✖ Estimate the cost of the project

4

Financing

etting a construction loan is a simple process, but just like with any other type of home loan it pays to shop around. The good news is that the process has become easier and the costs have become lower in recent years. Additionally, there are more choices in loans now than there were 10 years ago.

You must be business-like when approaching a lender. If you put yourself on the lender's side of the table for a moment, you can see that the lender is taking on a risk to loan you money — so you must convince him that the loan is not too much of a risk. The more you can convince him that you are a low risk, the more likely you are to get a loan. Convincing the lender can be done by preparing for the application. Do this by getting pre-qualified for a loan, and preparing and presenting a complete, well-documented application package.

The risk the lender is taking falls into several categories. The first risk category is the normal risk that a lender takes on every time he makes a loan. Are you financially sound? Are you likely to pay the loan payments? This category is where your credit score and financial qualifications will make him surer of you. If he sees a strong financial history, he will feel that you are more likely to repay the loan.

In addition to the financial responsibility, the building project itself is a risk to the lender, especially if you have never built before. Will the home meet code? Will it resell? Can you keep the project within budget? Can you get it done in a reasonable amount of time?

These are all questions that are valid. Lenders know that custom homes vacuum up money, and that most of the time they go over budget. So to satisfy the lender, you must be willing to take some extra steps and approach him with either a proposal or a portfolio that answers all of the questions he might ask. You need to convince him that your loan will not be a risk. Show him that you are not a builder who will leave any of the outcome to chance.

PRE-QUALIFICATION

The best way to be sure you will qualify for the loan before you spend a great deal of time and money on plans is to get pre-qualified. It is easy to complete the prequalification process, and it is free. When you are finished you will know for sure that the loan funds you intend to request are within your grasp.

The term prequalified ought to be called find out what you can borrow. By taking this extra step, you will know right away whether you can afford the 3,500 square foot home you were contemplating. Pre-qualification is fairly informal, but it will give you the confidence you need to prepare for the real application.

A lender will need certain forms. These may vary, but you will need to complete an application, a description of materials, and a construction cost breakdown. The application requires your employer information, assets and liabilities (including account numbers, balances, and the address of your banks), and income information (W-2s or tax returns). The lender can quickly tell you how much of a loan you will be able to get.

SHOP AROUND

After getting prequalified, look for the loan and the lender you will use for your construction process. It takes time to find a lender, and every lender has different requirements, so start at least several weeks ahead of the date that you want to begin construction. Check with the lender you select to find out what they want to see in your application. It pays to look around — especially since after reading this book you will be the most prepared applicant your lender has ever met. You will be considered a good candidate — so good the lender will want to negotiate terms for you.

YOUR SHOPPING LIST

The things you want to know when you get a construction loan are primarily the same as when you got a mortgage loan. You need to know how much the origination fee is, what the interest rate is, and whether the rate is adjustable or fixed. Can you lock in the rate? For how long? What will it cost? Is a down payment required? How much? What will your other costs be?

LOANS AND POINTS

A loan has two types of fees: the interest rate, which is figured over the life of the loan, and the points, which are up front fees that can offset the interest rate. Just as you negotiated on the price of your land, you can negotiate points and interest somewhat. This means you can pay a little more in points, in effect "buying down" your interest rate.

A point is equal to 1 percent of the loan amount. There are virtually no construction loans on the market for zero points. You should expect conventional lenders to offer a loan with two or three points. Generally if you offer to pay more points (say, three instead of two), you will have an interest reduction of one point. The loan officer can help you determine which situation is best for you.

WHICH LOAN IS BETTER?

Often when you compare two loans, the terms are different. So it seems impossible to compare them. There are online calculators that will compare the terms of two loans — type "compare two loans" into a search engine. The following outlines another way to compare.

Imagine you are borrowing $200,000. One bank offers you two points and 5 percent; the other wants three points, but the rate is only 3.75 percent. To compare them, first figure how much you will pay in points. Next, multiply the loan by 0.6 (for the construction portion of the loan) and calculate the interest.

Loan 1:

> 200,000 X .02 = 4,000
> 200,000 X .6 = 120,000
> 120,000 X 0.05 = 6,000
> 4,000 + 6,000=10,000

Loan 2:

> 200,000 X .03 = 6,000
> 120,000 X 0.0375 = 4,500
> 6,000 + 4,500=10,500

DRAWS

Draws are one of the characteristics of construction loans that are different from other loans you may have had in the past. When you take out a construction loan, you are able to draw (pay yourself the funds) at periodic intervals. You will need to know what the draw procedures are. Is an inspection required? Is notice required? Is there a fee? How often are you able to get a draw? How will they go about releasing the lien?

TYPES OF LOANS

When getting a loan, many people wonder if they are shopping for the loan and its terms or if they are shopping for a lender. In a sense, you are doing both. Hopefully, the lender who gives you the best service and listens to your needs will also be the one who offers the best loan package. At any rate, you will be better off if you shop around for the loan.

Should you use the lender you used the last time you purchased a home? Probably not, if you did not build it yourself. Construction loans are different from conventional purchase loans. If you can, find an experienced construction lender. Typical local banks may not have a construction loan department. You can check out their loan officers by asking them questions about the differences in construction loans and other kinds of loans. Ask why the fees vary and why certain forms are required; ask whether the draw reimbursement is preferable to a voucher system. You will soon be able to determine whether he or she knows construction loans.

There are several different kinds of loans and you can find a lender or mortgage broker who will listen to your needs and present you with honest answers rather than trying to sell you on the loan package of the week.

Land Loan

The first loan will be for the lot you plan to build on. After negotiating the price with the seller and getting a sales agreement in writing, it is time to consider the financing options available. If the seller will act as the lender, referred to as "carrying back paper," this may help you out over the short term, but when the bank funds the construction loan they will require that the seller be paid off. Normally seller financing does not give a better deal, and in many cases sellers charge as much as one-half to one point more in interest than banks. A seller may want to carry the loan to sell quickly or to put off paying the taxes on the sale of the property.

Occasionally, a seller offers financing because he knows banks will not lend money on the property. Sometimes this can be good, and sometimes it is because he is being deceptive. I ran into this problem myself a few years ago. A seller had 32 acres of lovely land and offered to carry the loan. As a self-employed person, I know that lenders are wary of people without a set income so I found the offer appealing. I signed a contract to purchase. A few days later I visited the property again and this time the neighbor stopped me. He said he owned the road and would move the road to another side of his own land so that I could not get to my property. Further investigation revealed that the neighbor was right; the acreage was landlocked. I spoke with county officials, who said that most people in this rural area would simply sell part of their land to give the neighbor access to the road, but in this case no one wanted to. I confronted the seller, who admitted he knew this. He refunded my money and canceled the contract — but this could have been a costly mistake. He could have held me to the purchase contract, and I would have owned land I could not build on or even get to. Watch out if the seller is too eager or insists on financing the land!

There may be other reasons property cannot be financed. If it is considered very large (over 20 acres), you may have to look harder to find a lender. If it has no electricity, it is already divided into multiple parcels, or there are buildings on it, you may not be able to obtain conventional lending.

The length and the rate are two of the most important ingredients when shopping for a land loan. The majority of lot loans cover a period of less than five years. If you believe you will get to the construction loan phase within two or three years, this should be fine. But if you feel that you will take a long time to design the house (or the local planning department is notorious for delays), it might be better to search for a 10-year land loan. Whether you choose a fixed or an adjustable rate may have less significance depending on how soon you are going to start building. If you are getting the construction loan fairly soon, it will pay off the land loan anyway.

Many buyers mistakenly believe that they must pay off the land loan before obtaining a construction loan. This is not quite the case. As mentioned in Chapter 2, you can do this using the equity you have in the land as leverage. Years ago it was something that people did as a matter of course; they bought the land, paid it off, and used it as collateral for the construction loan. You can if you want to, but these days a construction loan replaces the existing land loan when the time comes. It can be better not to pay off the land because you will need the money to pay the architect, designer, or contractor, and perhaps to pay for many permits. In addition, the interest you are paying on the land loan, although it is not much, is tax deductible. Better to keep the land loan and leave the cash in your pocket.

Construction Only Loan

This is the most common and the oldest type of construction loan. It is a short-term loan that only covers the construction period, which will be agreed on with your bank as some period from six to 18 months. As soon as construction is complete, you must arrange for another loan as the permanent financing. This not only means you must apply and qualify for another loan, it also involves all the loan fees and costs you paid before.

Many things can change while you are building the home. Will it be finished at the end of the construction loan term? Will the market or rate change significantly? Will the house qualify?

Banks are not terribly sympathetic when the 12 months are up and an owner-builder is unable to afford the new loan. For this reason, it is better to go with the permanent construction loan, also referred to as a single close or a one-time close.

No-Income Qualifier

You may hear about loans called No Doc, EZ Qualifier, etc. The gist of it is that the loans are marketed as requiring less documentation than other loans. For example, you might be able to state your income and assets without providing verification of them. The idea appeals to people who do not want to disclose their income for any reason, as well as people who are self-employed. These loans might require better credit and, in the end, may cost more than the documented methods of borrowing money.

You will need to provide a copy of the deed or the purchase contract, plus you will need the plans, materials description, and the cost breakdown. You might also be required to provide documentation of assets.

Construction-Permanent or Permanent Loan

These single-close loans were created in the 1990s, when banks finally realized that owner-builders were one of their best types of customer. After all, if you put all that work into building your home, are you going to lose it to default if you can help it?

The permanent construction loan rolls over into a long-term mortgage after construction is complete. This means you do not go through another qualification process or pay to have the property re-appraised. You do not pay for a second closing. Indy Mac even offers a package where you do not make payments during construction. Instead, you create an interest reserve, used to pay the interest that is accruing on the loan.

Paperwork the Bank Will Need

The residential loan application requires an incredible amount of paperwork. Fortunately, the lender will help you fill it out. If the permanent loan follows a construction loan from a different lender, you will need all the paperwork from your construction loan. Expect to provide:

✂ A complete cost breakdown (see following sample)

✂ A builder's statement if you are working with one

✂ A copy of the builder's license, if it is required in your state

✂ Copies of insurance policies, including builder's risk, flood insurance, liability, and worker's compensation (if appropriate)

SAMPLE COST BREAKDOWN FOR BANK LOAN APPLICATION				
Loan Information				
Borrower Name:				
Phone #:				
Property Address:				
Line Item Description:				
A. Pre Construction Costs:	Total Project Costs	Borrower Prepaid Costs	Changes to Budget	Remaining Funds
1. Architect, Engineering & Soils Study Fees				
2. Design Review / Plan Check Fees				
3. Permits - City / County				
4. Utility Connection Fees				
5. School / Park / Misc. Taxes				
6. Project Bonds				
Total Pre-Construction				
B. Construction Costs:	Total Project Costs	Borrower Prepaid Costs	Changes to Budget	Remaining Funds
General Requirements				
7. Temporary Utilities & Facilities				
8. Special Inspections / Testing- Geo-Tech, Structural				

SAMPLE COST BREAKDOWN FOR BANK LOAN APPLICATION

	Total Project Costs	Borrower Prepaid Costs	Changes to Budget	Remaining Funds
9. Job Security				
10. Equipment Rental				
11. Job site overhead				
12. Project Management / Supervision				
13. General Contractor's office overhead / profit				
14. State Sales Tax (where applicable)				
15. Builder Contingency				
Sub-Total General Requirements				
Site Preparation	Total Project Costs	Borrower Prepaid Costs	Changes to Budget	Remaining Funds
16. Demolition				
17. Clearing / Stakeout				
18. Rough grading / shoring / excavation / fill				
19. Site retaining walls / waterproofing / backfill				
20. Site drainage				
21. Private septic system				
22. Domestic Water Well				
23. Pump house & Pressure water system				
24. Environmental				
25. Off-site Improvements				
Sub-Total Site Preparation				
Foundation Complete with + A15 Foundation Endorsement	Total Project Costs	Borrower Prepaid Costs	Changes to Budget	Remaining Funds
26. Embedded hardware				
27. Ground Plumbing				
28. Ground Mechanical				
29. Ground Electrical				
30. Underground utilities				

SAMPLE COST BREAKDOWN FOR BANK LOAN APPLICATION

31. Foundation & Building retaining walls poured				
32. Concrete slab poured-house, garage				
Sub-Total Foundation Complete				
Building Rough-in Completion	Total Project Costs	Borrower Prepaid Costs	Changes to Budget	Remaining Funds
33. Structural masonry				
34. Rough framing materials				
35. Structural steel				
36. Modular or Sectional Mfg. Home				
37. Package / Kit Home				
38. A51 Trusses / components				
39. Rough framing labor				
40. Lightweight concrete interior floors				
41. Plumbing top-out				
42. Rough heating, ventilation, air conditioning				
43. Rough electrical				
44. Fire protection - sprinklers				
45. Fireplaces and Flues				
46. Security & Communications pre-wiring				
Sub-Total Building Rough-in Completion				
Exterior Weather-tight	Total Project Costs	Borrower Prepaid Costs	Changes to Budget	Remaining Funds
47. Waterproofing decks, etc.				
48. Sheet metal, Gutters, downspouts				
49. Roof Covering				
50. Windows				
51. Exterior Doors				

SAMPLE COST BREAKDOWN FOR BANK LOAN APPLICATION

52. Skylights				
53. Glazing				
54. Exterior Siding				
55. Exterior Trim				
56. Stucco				
57. Masonry veneer				
58. Ornamental Iron				
59. Garage Doors				
60. Exterior Painting				
Sub-Total Exterior + A55				
Drywall/Finish Carpentry	Total Project Costs	Borrower Prepaid Costs	Changes to Budget	Remaining Funds
61. Insulation				
62. Drywall/Plaster				
63. Interior Stairways				
64. Cabinetry				
65. Finish Materials / Millwork				
66. Interior Doors				
67. Finish Hardware				
68. Finish Carpentry Labor				
Sub-Total Drywall / Finish Carpentry				
Building Completion / Final Inspection	Total Project Costs	Borrower Prepaid Costs	Changes to Budget	Remaining Funds
69. Countertops				
70. Tub / shower / enclosures				
71. Interior painting / Wall coverings				
72. Hard surface finish flooring				
73. Carpeting				
74. Built-in Appliances				
75. Special Equipment				
76. Security System				

SAMPLE COST BREAKDOWN FOR BANK LOAN APPLICATION				
77. Intercom				
78. Built-in Vacuum Cleaner				
79. Finish Plumbing				
80. Plumbing Fixtures				
81. Finish Electrical				
82. Lighting Fixtures				
83. Finish Heating, Ventilating, Air Conditioning				
84. Solar Backup				
85. Bath Accessories				
86. Tub, Shower Doors, Mirrors				
87. Finish Grading				
88. Pool / Spa				
89. Hardscape (Driveway, Walkways, Steps)				
90. Landscaping				
91. Irrigation System				
92. Fencing and _A90 Gates				
93. Touch-up / Final Cleaning				
Sub-Total Building Completion				
Total Construction Costs				
Total Line Item Cost Breakdown				

WHEN TO APPLY

You may be wondering when to apply for your construction loan. After all, the loan documentation, like your credit report and appraisal, will expire after 90 days. So if you apply and then do not fund the loan within 90 days, you will have to pay more fees and reapply. On the other hand, if you have to wait around on the financing, you may waste valuable construction time or lose one or more of the subs you had lined up.

Most building departments take one to two months to issue permits. So, if you apply for the loan about two months before you believe you will break ground, you will have the funding at exactly the right time.

If you have already started building before you apply for the loan, you can get it funded easily. The bank may require that you provide them with title insurance and a deed of trust.

TRUTH IN LENDING

Because of the Federal Truth in Lending Act, people who want to borrow money to buy or build real estate have certain rights. Article Z of this Act requires that the lender disclose to you the details of the loan he or she is offering you, including the interest rate, the terms, the extra costs, and variable rate features in the APR.

If the lender does not disclose the APR of the mortgage loan to you, then you are able to receive your application fee back as a refund. You will normally receive all the disclosures in writing when you make the loan application. If any of the terms change before closing and you decide not to proceed with the loan, the lender must return all fees to you.

After closing, you have three days to change your mind. If you make a request in writing to the lender, all your fees will be refunded.

DIFFICULTIES

There can be difficulties when applying for a construction loan. Most of these are items you can get around if you are prepared for them ahead of time. The best way to prepare is to check and double-check on financing, the house plans, and the process. Be prepared to be as flexible as possible throughout the project.

CREDIT

Your credit may be one of the first things that can get in the way of your construction project. Credit is based on a scoring system called the FICO score. Poor credit is considered to be about 300; average is 620; and good credit is 660 or better. Scores go up as high as 850.

Credit scoring is based on your pattern of making payments (35 percent), how much money you owe (30 percent) and the amount of available credit. Lenders base their loan approval, and possibly the interest rate, on your credit score. If your credit score is below 620, there may be problems getting a loan through conventional banks — just like any other type of loan. So if you know your score is low, you might approach non-conventional lenders.

The cost of bad credit is high when you are talking about permanent loans. If your score is 650 and the rate offered to you is one point higher than the one offered to others with better credit, you could end up paying $21,000 more over the life of a $170,000 loan.

Credit reports often contain inaccuracies — and sometimes they have information that does not belong to you. As soon as you begin thinking about building a home, get a copy of your credit report. These are free once a year from each of the three credit reporting agencies. You can also request a free annual report at **www.annualcreditreport.com** or by phoning 1-877-322-8228. Once you have a copy of your report, study it carefully.

If there are problems or false information, contact the agency right away. You can do this by mail or via the Internet. Give them supporting documentation of the item in dispute. The complaint will be investigated within 30 days; they are resolved quickly.

You can get a copy for free from any of the "Big 3" credit agencies. Contact them at:

Equifax
PO Box 740241
Atlanta, GA 30374-0241
800-685-1111

Experian
P.O. Box 2002
Allen, TX 75013
888-397-3742

TransUnion
P.O. Box 390
Springfield, PA 19064
800-916-8800

UNEXPECTED FEES

When building a home, you will feel as if the array of fees is endless. The best way to prevent surprise is to ask and re-ask the lender, the local government agency, and your subcontractors and suppliers about fees. But beyond that, there will still be unexpected charges. This is where strong cash reserves will come in handy.

One of the big surprises, especially if you are already familiar with purchasing real estate, is the appraisal. Construction loan appraisals require more time and more cost than the average purchase loan appraisal, so the extra expense is passed on to you. If you are borrowing a lot of money, expect the cost of appraisal to be especially high — perhaps double what you expected. Additionally, an appraisal only lasts 90 days, and you will be charged $150-$250 for the privilege of having it updated if you need to. And if you change lenders, you will pay a fee to have the appraisal retyped, if the appraiser agrees with it. Often they do not, and you will have to order a completely new appraisal.

Junk fees are another area where you can get hit with surprises. These are the small fees charged by various parties in your construction loan. Lenders will break this down for you in writing. Despite the name, most of the costs are necessary. There may be inspection fees, administration fees, wire transfer or messenger fees, processing, recording, certifications, and more. Get a good faith estimate in writing up front and examine the list carefully — but most of the fees are for services that must be paid for in order to process the loan.

PASSING MUSTER

Often, the loan committee at the bank will need to see qualifications for building a house that you do not have. In this case, you could call around and try to find a general contractor who will provide a copy of his license for the files. This seems to satisfy the requirement as far as the lender is concerned. You might try to find a semi-retired contractor to do this — or even offer a fee. If the bank asks that the contractor sign an agreement to perform services, let him select an hourly rate for consulting. Even if no money changes hands, this seems to be satisfactory. Again, this is just a formality — the bank wants to see that the money they are lending is going for a house that will have value in the end.

BEING TOO CONSERVATIVE ON THE TIME FRAME

If you get a construction-only loan, it will be offered for somewhere from six to 18 months. Many people mistakenly choose the short term, only to be surprised later. The shorter-term loans seem attractive, since they carry a better interest rate. But if you extend past the loan term, you could end up paying huge penalties — maybe as high as 0.5 percent of the loan amount (not the interest, the total loan). Extending past the loan term can end up being quite costly.

Many people set a goal of six months, and some use a goal of four months for completion. In theory, there is no reason you cannot complete it in that amount of time. But in reality, you do not know the system yet. You most likely do not know the inspectors, the government entities, or the subs. There is a chance that there could be delays during the construction process.

To avoid all the stress and extra expense, try to figure the length of time you believe the home will be under construction, then double it and use that as the loan term. This should allow enough of a cushion to offset delays, weather problems, losing subs, etc. The extra percentage will be very little compared to the tension people feel when the loan term is nearly up.

BUDGET WILL NOT WORK WITH YOUR LOAN ABILITIES

Not talking with lenders until after the permit process can get you into a lot of trouble. All lenders have certain guidelines, and if your budget does not fit you may not qualify for a loan. This can delay the project for years. Find out what you can afford to build first, then create a plan.

CHECKLIST FOR CONSTRUCTION LOAN DOCUMENTS

⚒ Application Form   ⚒ Consent Form

⚒ Signed Disclosure   ⚒ Most Recent Pay Stubs

⚒ Other Asset Records   ⚒ Land Contract or Deed

⚒ Survey   ⚒ Plans — 3 sets

⚒ Specifications   ⚒ Cost Breakdown

- Title Insurance
- Construction Insurance
- Architect Information
- 2 Years of W-2s (or last 2 years' tax returns if self employed)
- Rental Agreements (if you own rental properties)
- Three Months of Bank Statements (on all accounts)
- Retirement Account Statements
- Loan Documents (for land with loan)
- Proof of Regulatory Compliance
- Builder's Contract
- Permits

Extra Documents to Make You Stand Out from the Crowd

- Copies of subcontractor and supplier bids
- Copy of your project timeline
- The bank's own lien release form, signed by subs (this is a form subcontractors sign saying they have been paid and will not put a lien on the property)
- Résumés of contractor or major subs you are using

CHAPTER 4 CHECKLIST

- Obtain your credit report

- Pre-qualify for the loan
- Investigate financing
- Understand points versus interest rate and how to calculate the best deal for you
- Understand different loan types and draws
- Shop around for a loan
- Select lender
- Complete checklist for construction loan types and draws
- Complete loan application
- Apply for loan
- Review potential problems
- Unexpected fees
- Budgeting
- Time frame for construction

Financing for Remodeling

FINANCING A REMODELING PROJECT

Most of this book has focused on building, but I know that many people are reading it to prepare for a renovation or to build an addition. If your remodeling or addition project is extensive, you may be wondering where you will get the money. It is much easier than most people realize to get financing for a home upgrade — especially if you have built up equity in the house.

The easiest and quickest way to get money to finance the project is through a second mortgage on the home. This can come in the form of a home equity loan or a home equity line of credit. The difference between the two is that with a loan you have a closing and receive all the money at once; with a line of credit you can make "draws" and take out the money as you need it, so you save a little in interest if you do not draw out the money until the moment it is actually going into use.

Your home serves as collateral on both types of loans. You will be approved for a certain limit that is based on the home's appraised value and the equity that you have in it. For example, if your home is worth $100,000, the credit limit the banks will allow will probably be 75 percent of that or $75,000. But if you still owe $35,000 on the home, you only have $40,000 in equity built up in the property. So the lender will loan you up to that amount — $40,000.

Many of the home equity line of credit plans are good for a specific period of time, like five years. During that time, you can draw on the money up to the agreed upon limit. At the end of that time, you cannot borrow additional funds unless you renew the loan. Some plans have a minimum draw or a minimum outstanding amount.

Most home equity lines of credit offer a variable interest rate. That means it changes with the prime rate and will be offered as that index plus a margin, like 2 percent. Variable rate plans must have a cap, or limit, on how much the interest rate can increase over the life of the loan. Be sure to find out how high your interest rate can go. Also find out whether the loan will allow you to convert the variable rate to a fixed rate should you decide to. The interest on home equity loans is tax deductible — it depends on your personal financial situation, so check with your tax advisor to be sure.

THE COST OF A HOME EQUITY LINE OF CREDIT

The costs for home equity loans involve an application fee, an appraisal fee, and some up front charges like one or more points. There will also be the normal sort of closing costs: attorney fees, mortgage preparation, taxes, property insurance, and title insurance. There may also be transaction fees and additional fees each time you draw money. These are all costs that should be revealed to you in writing when you apply for a loan.

THE COST OF A SECOND MORTGAGE

Compared to a line of credit, a second mortgage loan may seem like a better deal. It will give you a fixed loan amount and it can be paid over a fixed time period. If you choose a fixed rate loan, all the payments will be equal. When comparing the loans, be sure to compare the closing costs and the interest rates — but the APR for a line of credit is figured differently. It is based on the periodic interest rate, whereas the APR for traditional mortgages is figured on the interest plus points and other finance charges.

APPLYING FOR THE LOAN

Just like a construction loan, a home equity loan will require a breakdown of labor and materials costs. Be sure to include all your permit fees and equipment rental, then add a cushion of at least 20 to 30 percent for unexpected expenses. Remodeling tends to have more cost overruns, in part because you do not realize certain problems that may be uncovered as you begin work. Hidden structural defects, extra wiring that needs to be replaced, or other surprises may cause more money (as well as more time) to be spent on the project.

The amount of money you can borrow will depend on the value of your home, the loan-to-value ratio, your income, and your credit rating. The house payment plus your other debt should fall below 36 percent of your gross monthly income to qualify for the loan. For credit that is less than perfect, you may need to pay points to get the loan.

SPECIAL LOAN PROGRAMS

When borrowing to remodel, there is an FHA-backed loan called an FHA 203(k) mortgage that will allow you to refinance your home and roll remodeling costs into a new loan. The loan is based on the estimated worth of the house after you have completed the improvements. The loan limits will vary by county; they tend to be somewhat low, but if you qualify you will seem to have more equity, and therefore the amount you can borrow will be higher than with some other loans.

Another kind of mortgage that is worth looking into is an Energy Efficient Mortgage, or EEM. This kind of loan can increase your debt-to-income ratio by 2 percent. Your home will have to meet rather stringent energy efficiency standards to qualify. Check with Fannie Mae for details.

B AND C LOANS

These are loans that work for people who do not have good credit or who do not fit the normal employment criteria. Lenders push them for debt consolidation, but they are available almost everywhere. They are non-conventional loans so the interest and fees may be higher. They are also difficult to compare because the requirements and terms vary.

OTHER COLLATERAL

If you are borrowing to improve your home and you have assets other than your house, you might use a different type of loan. Stocks, bonds, savings, retirement accounts, and CDs can be used as loan collateral. There would not be the closing costs that would be expected on a second mortgage or home equity line of credit. Of course, you will not have tax-deductible interest either. But the rate might be low enough to justify all that; it is certainly worth checking into.

Obtaining financing for a remodeling project is much easier than getting financing to build a new home. It is also much faster. Just be sure that you check into all the options available to you, and that you understand all the terms of the contract before you sign.

CHAPTER 5 CHECKLIST

- Investigate Costs and Availability of Home Equity (HELOC) versus 2nd Mortgage Loans

✖ Investigate Other Available Types of Loans

✖ Government Loans (such as FHA, EEM)

✖ B and C loans

✖ Collateral Loans (other than your home)

✖ Private Loans

✖ Determine the Best Type of Loan for Your Situation

Insurance

OBTAINING ESSENTIAL INSURANCE INFORMATION

There are several kinds of insurance that will be necessary throughout the purchase and building process. Whether you are working with a lender who requires it or financing your project 100 percent yourself, you will still want protection. Not all builders' policies are covered by every agency, so give yourself plenty of time to find the kind of insurance you need. Some of the more common kinds of necessary insurance policies are described in this chapter.

TITLE INSURANCE

This is a policy that you obtain when you purchase the land. The seller should ensure the property has clear title (free of defects, liens, and encumbrances unless they are listed in the policy). Most property in the United States has gone through ownership changes, and there could be a weak link anywhere along the way. For example, there could be liens, unpaid taxes, or fraud. Title insurance covers you in case of any of these claims. Mortgage lenders require borrowers to carry it. The insurance is purchased with a one-time payment made on the front end of the loan. It is good until the loan is repaid. It protects you and your heirs indefinitely as long as you have an interest in the property. It will also cover loss and damages if the title is unmarketable (a legal term meaning having defect or likely to end up in litigation) or if there is not a right of access to the land. A title insurance policy protects you from:

- Errors or omissions in the deed
- Liens for unpaid taxes.
- Liens by contractors
- Recording mistakes
- Undisclosed heirs
- Forgery

If there is a lien against you during the time you own the property, your title policy will not cover you. You will be required to take steps to have the lien removed prior to selling the property. If you refinance the loan, you do not have to get a new owner's policy, but the lender will require a new lender policy. Even if you are just refinancing with the same lender, the title insurance policy terminates when you pay off the mortgage.

Title insurance rates are not regulated in Alabama, The District of Columbia, Georgia, Hawaii, Illinois, Indiana, Massachusetts, Oklahoma, and West Virginia. In Texas and New Mexico, the state sets the price. In other states, it may pay to shop for the policy.

Money-Saving Tip:

If the property you are buying has been sold within the past five years, ask the title insurance company for a discount. You might also be eligible for a discount if you are a first-time buyer.

GENERAL LIABILITY INSURANCE

As an owner-builder, the lender will require you to carry a minimum amount for property and personal injury, $300,000 or $500,000. It can be a comprehensive general policy or, in some cases, it may be a broad form liability endorsement. Does this insurance seem like a waste of money to you? Consider this: In 1970, Congress passed the Occupational Safety and Health Act. The wording of this act in part reads "to assure so far as possible every working man and woman in the nation safe and healthful working conditions and to preserve our human resources." This relates directly to construction, where fatalities commonly happen due to falling from a roof or ladder, being struck by an object or by equipment, and by receiving an electrical shock. So it is vital that you carry liability insurance on your construction project.

Liability coverage protects the insured for damages that result from third party claims. These could be bodily injury, property damage, advertising injury and personal injury. General liability covers both premises operations and the products completed, like construction defects.

It is of the utmost importance that you go over the liability policy and understand exactly what it covers. Some policies may exclude theft, wear and tear, machinery, testing, workmanship or materials, design error, collapse, flood, earthquake, and the like.

WORKER'S COMPENSATION INSURANCE

If you have employees, you will carry a worker's compensation policy. If you hire a contractor, he will carry the policy — unless all the labor and subs are independent contractors. If they are, then they will fall under your liability policy if there is an accident. Lenders may require that you and the contractor, if you have one, sign a waiver stating that the lender is not liable for workers' compensation violations.

People who employ family members or who hire a licensed contractor to oversee the project do not have to carry workers' comp. You also do not have to have it if you use independent subcontractors. However, if you employ anyone else in any capacity or if you participate in federal or state withholding for someone who works for you, then you may be required to carry workers' compensation insurance. Check on the IRS Web site at **www.irs.gov** in the section "Who is an employer?" to be sure.

Your subcontractors will have workers' compensation for their own employees. They will also have Builder's Liability Insurance, which they should show you. In addition to asking for their proof of insurance, you should also ask your subs for evidence of their license and bond.

COURSE OF CONSTRUCTION INSURANCE

The course of construction insurance policy might also be called builder's risk. Policies of this type vary so it pays to check around. The policy should be in force before closing the construction loan or before materials arrive at the job site. The lender will be the payee in the policy.

This is a policy that protects against theft, vandalism, fire, weather, and other kinds of damage during construction. It is sometimes referred to as an all risk policy because of its vast coverage. If some strange accident occurs, the policy will pay for the cost of rebuilding the home back to the way it was prior to the accident, up to the loan amount. The lender will require the policy.

OTHER INSURANCE

If your home has been designated as a Special Flood Hazard area, you must carry flood insurance. Aside from this, check with your insurance agent to see what kinds of policies are appropriate for your situation and locale.

Sometimes people purchase dwelling and fire insurance instead of construction insurance. After construction is complete, the fire policy can be converted to homeowner's insurance at a savings.

HOMEOWNER'S INSURANCE

Once the home is complete, it is time to convert the fire policy or purchase a new homeowner's insurance policy. Be sure that you know the true value of the home, and that your policy will cover the total cost of replacement, should that ever be necessary.

WHAT IF YOUR CONSTRUCTION PROCESS DAMAGES THE NEIGHBOR'S PROPERTY?

During the course of construction, the neighbor may sustain damage. This could be from a tree falling onto their fence or other property. It might be due to a subcontractor or supplier driving onto the neighbor's lawn. It could be one of many other accidental scenarios.

Normally if a tree falls and it is not due to negligence, the neighbors' own insurance policy will cover the damage. If you are negligent, though, your neighbor could file a lawsuit and your own insurance would come into play.

THINGS YOU CAN DO TO ENSURE SAFETY ON THE JOB SITE

Just as OSHA requires builders to create safety plans, so should you as an owner-builder. Safety should be your number one concern — not only to protect you and your investment, but also to protect the health and well-being of your workers.

You can keep safety first by operating just like a general contractor would. Safety meetings, referred to in the trade as tailgate meetings, are common. They are informal, short discussions about keeping the work area safe from injury, fire, and other hazards. Utilize the following checklist to monitor safety on your site:

✖ Arrange for a portable toilet on site.

✖ Provide fresh drinking water and disposable cups.

✖ Post Safety Rules onsite and also communicate them in writing or verbally to each trade contractor.

✖ Keep a first-aid kit available.

✖ Post the phone numbers for police, ambulance, and fire station. Consider installing a telephone for this purpose.

✖ Ground the temporary electrical service and all electrical tools.

✖ Check often to be sure electrical cords are kept away from water.

✖ Use only equipment that is listed, labeled, or certified. Use in accordance with manufacturer's instructions.

✖ Post Warning and Danger signs as appropriate.

✖ Insist on hard hats and steel-toed boots as appropriate.

✖ Cap protruding steel rebar, nails, and other protruding dangers.

✖ Ensure that power tools are well-maintained and properly stored.

✖ Provide protective gear — goggles, gloves, and respirators.

✖ Require protective goggles when eye injuries are possible.

✖ Require wearing of personal protective equipment.

✖ Set a good example.

✖ Ensure there is adequate slope and proper fencing on edges of all ditches and trenches over 4' deep.

✖ Place excavated material at least 2' from edge of ditches and trenches.

✖ Cover open holes in sub-floors.

✖ Safe access to and use of all types of scaffolds.

✖ Guardrails installed on all open-sided floors or platforms.

✖ Workers on roof only with proper equipment.

✖ Stair rail system constructed on stairways of four or more risers.

✖ Excess and/or flammable scrap removed daily.

✖ Use only approved containers for storing flammable or combustible liquids.

✖ Gas cans and other flammable liquids must remain in secure area.

✖ Welding tanks must be shut off tightly when not in use — check and re-check.

✖ Monitor area where soldering work is done — look for smoldering or burning wood.

✖ Insist that all workers maintain proper clearance from power lines.

✖ Spread oily or paint rags outside to dry before disposal so they will not ignite.

✖ Keep a Material Safety Data Sheet on site for any hazardous chemicals that may be encountered.

✖ Frequent, daily safety checks and clean-ups are most effective and will remind workers to be safety conscious.

OFFER SUPPLIES ON-SITE

One experienced builder uses a storage system of bins for items like nails, screws, and small tools that most of the subcontractors will need. He feels that the expense of filling and supplying the bins more than makes up for the trouble, because of the amount of time saved. Workers do not waste time running to the lumber supply store to get something they need.

Owner-builders should incorporate this system into the project they build. Simply fasten a bunch of plastic storage containers together, or attach them onto a framework of lumber. Locate them centrally on the site, and fill them with the basic equipment needed. Make these available to the work crews when they come on site. You will find that the site remains much cleaner, with fewer scattered items lying around.

SAFETY INSPECTIONS

Throughout the project, you will encounter inspectors. The various inspections are covered in detail in Chapter 8. There are county and local officials, as well as inspectors who are sent by the lenders. These folks are not the enemy. They are there to see that your work meets the structural standards and safety requirements that it should. If the work fails inspection, you will be given time to correct it. You will then reschedule the inspection. You may be asked to pay an extra fee for the repeat inspection.

County and local officials will normally stick to a set schedule of required inspections. For example, they may want to see the property before pouring the concrete footings, before pouring the concrete foundation, after framing, after the electrical/HVAC/plumbing rough-in, after laying the sewer line, and after the project is complete.

To meet these inspectors with the best attitude, there are a few things you can do to be prepared for them to show up. First, find out how much notice the inspectors will need. You should find this out early, when you apply for the building permit.

Once you know when to notify the inspectors, always give them advance notice that you are ready for inspection. If you happen to make the appointment and then find that for some reason you are not ready, be sure to give them a call so that they can reschedule.

On the day of the inspection, make sure the site is clean and organized. Be present during the inspection so that if there is something you must correct you can learn exactly what it is and what steps you need to take.

Have a good attitude, even if you are asked to change something — the inspectors are only doing their job. You may even be able to learn something from them.

CHAPTER 6 CHECKLIST

- Understand the different types of Insurance needed and obtain
    - Title Insurance
    - General Liability Insurance
    - Workers' Compensation Insurance
    - Course of Construction Insurance
    - Flood Insurance
    - Homeowner's Insurance
- Complete jobsite safety checklist
- Conduct regular jobsite safety talks and inspections

7

Find & Negotiate with Subcontractors

A subcontractor is someone you hire to complete all or part of your project. Most of the subs you will work with are involved in specialty trades; you might hire one person to build a block foundation, another to pour the concrete, and another to waterproof the basement.

THE IMPORTANCE OF THE SUBCONTRACTOR

The subcontractor's abilities and talents in his trade (or lack of them) will make or break that particular aspect of your home. If someone does a bad job you will have to look at that every day you live in the house. I remember a house I nearly made an offer on years ago. The wall in the bathroom had been poorly finished, resulting in an uneven finish and a crooked, thick corner instead of a sharp one. I asked the builder whether he could get the drywall contractor to come back and smooth the work. It should have been a one- or two-hour job to sand and respackle it. He instantly said no. Knowing what I do now, I suspect there had been more problems with the sub than just the bathroom walls. Because of the short, sharp answer, I thought the contractor was not cooperative; I let the deal pass by. The contractor lost a sale because of the subcontractor's workmanship.

Money is at the front of everyone's mind when building a home, but choosing the right sub is more complicated than looking for the lowest price. Prices have a way of creeping up, even though they never creep down! Most people focus too much on choosing the lowest bid, and that is what gets them into trouble. The right contractor will produce quality work, be reliable and show up when he says he will, and offer a reasonable price. He will not necessarily be the lowest bidder; in fact, he probably will not be.

WHERE TO FIND POTENTIAL SUBCONTRACTORS

Locating a good sub is difficult — and finding a poor sub is easy. Those who do good work and are reliable are often busy, so be sure to start looking well ahead of the date you will actually need them.

Local Building Supply

The easiest place to find the names of subcontractors is at your local supplier. These people do business with everyone so they can readily tell you who is staying busy. There are often bulletin boards where various workers can place their business cards. By visiting the supplier and talking with employees and contractors, you can pick up the names of several people who are staying busy at their trade.

Your Carpenter

The lead carpenter you hire can be a wealth of information on other subcontractors. Carpenters are present on the job through more of the process than other subs, and as a result they know all the other professionals. Trust your carpenter to give you the names of reliable workers.

Visit Some Job Sites

Start looking for subs through your local building supplier or the carpenter you have hired. If you cannot find one this way, try visiting a few job sites. Some will have signs advertising the various subs; if not, simply find a house that is currently under construction, stop, and ask questions. The subcontractors on the site will most likely give you their contact information, references, and maybe even a price, all in a few minutes' time. If the boss is present on the job, consider asking him, too — far from being upset that you are there, he can offer valuable insight. Be sure to get his name, so that if you do get in a bind or decide to hire someone to oversee this or a later project you can get in touch with him.

The Yellow Pages

Traditionally, the yellow pages are not the best place to find subs. For one thing, many carpenters who work as independents are not listed. There are also other trades that will not be listed. The yellow pages do list most heating and air companies, plumbers, lumber suppliers, and electricians.

THE BIDDING PROCESS

When gathering bids, be sure that you get at least three or four on each job you want to subcontract. My policy is this:

- Every bid must be detailed in writin.
- Every sub must agree to sign a written contract.
- Every bid is for the job, not by the hour.

STEP-BY-STEP THROUGH THE BID PROCESS

Keep names of potential subs in separate sections in your project notebook. When you have made all the design and material decisions (as best you can, since some of this will require input from the sub) prepare a written set of specifications for the job.

Now set up a file for each sub that you will be asking to bid. This file should contain a copy of your specs, a copy of your contract, and the bid sheet you want submitted to you. Some of your subs will have their own computer software to create bids or they may simply have another form they prefer to use. There is nothing wrong with this; just be sure that you have their workmen's compensation details and that there are signatures on the agreement you decide to use.

Following is a sample bid sheet. When you receive the bids, the most important item to consider is how detailed the specifications are. If they are not detailed, you will need to go back to the contractor and ask questions to get a complete, accurate bid.

While you are creating files, set up a file of your own to keep track of bids. This one may be useful, although yours may have more or less categories:

SUB	NAME	BID	GOOD UNTIL	NAME	BID	GOOD UNTIL	NAME	BID	GOOD UNTIL	BEST BID
Excavation										
Concrete										
Framing										
Roofing										
Plumbing										
HVAC										
Electrical										
Masonry										
Siding										
Insulation										
Drywall										
Trim										
Paint										
Carpentry										
Flooring										
Tile										

SUB	NAME	BID	GOOD UNTIL	NAME	BID	GOOD UNTIL	NAME	BID	GOOD UNTIL	BEST BID
Glass										
Gutters										
Asphalt										

SAMPLE SUBCONTRACTOR BID SHEET

Name: _____

Address: _____

Address of job that is up for bid (fill in your address): _____
_____

Date: _____

Amount of Bid: _____

Workmen's Compensation Insurance Company Name: _____

Agent's Name:_____

Certificate Number: _____ Expiration Date:_____

(attach copy of insurance card)

References (list name, address, and telephone contact number for each reference):

1. _____

2. _____

3. _____

Work to be performed: _____

Materials you will supply: _____

Materials homeowner will supply: _____

Signature: _____ Date: _____

Signature: _____ Date: _____

Once you have gathered all the bids, discuss your plans, including your deadlines, with each potential subcontractor. Make sure that by the end of the conversation you are clear about who will purchase materials, how the contract will be executed, and how payments are going to be made. If the written bid does not give a breakdown of exactly what materials are going to be used, find out.

QUESTIONS FOR SUBS

It is useful to ask questions of a sub, both to get to know him (after all, he will be your employee) and to help you make an educated decision. However, if you have never built or remodeled before, you may be at a loss as to what questions to ask. Besides the usual questions, consider these:

- Do you do the work or do you send a crew? How big a crew would you send? How experienced are they?

- Why should I hire you over the other subcontractors?

- What would you do if this were your own house?

- What are some ways I can save money?

- Is there any thing I can do to save labor costs?

- What other contractors should I coordinate this job with? Is there someone you like or recommend to do that job?

- How far ahead should I schedule you?

- How long will it take you to do the job?

- What other suggestions do you have for me?

After you have talked to several subcontractors, you might go back and ask more questions:

✘ Why is your bid so much higher (or lower) than everyone else's?

✘ What will I get for the extra cost?

✘ How can I save money?

One owner-builder cautions that you should not hurry after you receive bids. "People want to jump on the first bid or two, because they are impatient. They want to get the job done now. Plus, if you are doing an extensive remodel and you need, say, eight subs, and you want at least three to five bids for each, that could be 40 bids."

That is okay. It does not matter if you have to create 40 sets of plans, or do 40 interviews in 30 days. Plan to spend the time that is necessary to get all the bids you need. Some builders allow the first month strictly for bid gathering, and the next month to revisit and negotiate each bid.

E-MAIL

Should you conduct part of your interviews by e-mail? This can certainly cut down on the trouble and expense of creating plans, because you can fax from your computer or send them as e-mail attachments. Plus, if you send out a request for bids with specific instructions in a fill-in-the-blank format, you can get lots of detailed bids quickly.

The disadvantage of using e-mail for part of your interview process is that by not being face-to-face, you lose out on some of the meaning of the conversation. However, the distinct advantage is that every discussion is recorded in writing; if a contractor does not follow through on what he promised, you have a written record of it.

HOW TO REALLY CHECK OUT YOUR CONTRACTORS

We have all heard stories of contractors who grab the cash and disappear without doing the work. It is almost impossible to avoid this, especially if you build or remodel homes often. But here are a few suggestions that might help you prevent such problems.

Visit construction sites or new homes and get the name of anyone whose work you like.

Put yourself in the path of the trade. Join the Home Builders Association and get to know the members. Go to home and garden shows and talk to everyone in each booth about your project.

Ask each sub for the names of other trade subs. In other words, ask the excavator to recommend a sub for the foundation; ask the framer who might be best for insulation.

Check every name that you get against the state registrar's list of contractors, as well as the Better Business Bureau. It is common for contractors to have complaints on file, especially if they have been in business for a long time. One complaint is not a red flag for a builder with a large volume of work. But if the volume is smaller and there are several complaints on file, consider moving that name further down your list or toss the name out altogether.

SELECTING THE RIGHT CONTRACTOR

Now that you have written bids in front of you and have talked individually with each subcontractor, you should have a fairly clear assessment of them. Select the top three bids. This choice should be based equally on price and on your concept of the subcontractor's character. Do not be afraid to eliminate someone if you have a bad feeling about him or her. Elana relates the story of her shady sub:

"I did not want to use this guy, but the contractor who built the home insisted that he was okay. I had bad vibes all over the place. When he showed up to replace the windows, the dog — who never barks — started barking. And barking. The whole time the guy was there, the dog barked at him. I caught him cussing at the dog and then kicking it. The reason I did not put the dog up is because it made me sure I was right about the man. I only let him replace the one window and then I fired him. Just the one took him two entire days. Later I found that the window was not installed properly; it was crooked inside the frame, and would not even open!"

After you have selected your final three subs, compare their bids against your original budget. It is amazing how variable bids are. You will find that some subcontractors underbid the job. For excessively low bids, try to determine whether they are unable to estimate properly, they are new to the business and just need a job, or they are frantic to get work. If one bid is more than 10 percent or so less than the others, and you cannot find a viable reason for it, eliminate that one. The same goes for extremely high bids. If the average bid is $6,500 and one is for $9,500 it is too far away from the norm.

Is there a difference in the average of the top three bids and the price you expected to pay? If so, choose the sub you consider to be the top prospect and ask him what his best cash price is. Some subs will not change the price for a cash payment, but for others, money talks. Many subs expect to negotiate the price by 5 percent or more, especially since so many books are on the market advising homeowners to ask them to lower the price.

When you are sure you have negotiated the price as far as you can go without sacrificing quality, select the sub. Let him know right away that you will need a commitment for the given time period that you need him. You should have already outlined the working schedule so that you can tell him when that date is.

CONTRACTUAL LANGUAGE

Terms and Conditions: All agreements and oral promises between you and each contractor should be put in writing. This is done to protect both the customer and the contractor. If you intend to do some of the work yourself or if you are hiring more than one contractor for jobs that would normally be done by the same person, those issues need to be entered in the contract as well. At the least, the written contract should include:

✖ A thorough description of the work to be done.

✖ A complete list of all materials to be used — quality, quantity, weight, color, size, brand name, and so on. Use part numbers whenever possible.

✖ The total cost, including a breakdown of charges for both labor and materials.

✖ The start and completion dates that you have agreed on.

✖ A payment schedule.

✖ A right-of-rescission clause, which states your right to cancel the contract within three business days. This clause should also state that the contract is null and void or can be renegotiated if the job uncovers unexpected or hidden problems or damage after the work has begun.

Warranty Clause: Any warranty offered on products or services by the contractor should be in writing. Study the warranty carefully; make certain you understand all the terms and conditions, including the length of the warranty. The warranty must state whether it is a full warranty that gives the consumer certain automatic rights or a limited warranty that restricts certain consumer rights.

In addition, make sure the written contract includes:

- The contractor's full name
- His address
- All telephone numbers (home, work, cell)
- A professional license number

Also, make sure that the contract discusses debris and material removal once the job is finished. Make sure you both understand who is going to remove it and who carries the cost.

Once you have a completed contract, go over it with each subcontractor, making changes together as needed. Never sign a partial or blank contract. Read every contract clause carefully and ask any questions you may have before signing. Both parties must initial any changes to the typed agreement. Get signatures and make copies; retain one for your office, an extra for yourself, and give a copy to the sub.

Be sure to stay in contact with your subcontractors, especially if it is going to be some time before they are needed. If the schedule happens to change (and it will) let each sub who is affected by the change know as soon as possible. This is the only house you are building, but it is not the only one your subcontractor is working on. By letting him know your plans, he can adjust his schedule and still get paid — and he is much more likely to be available when you are ready for him.

PAYING SUBCONTRACTORS

When the written contract is hammered out between you and the sub, it will contain a price for the work. This may be a certain number of dollars per square foot or a lump sum. While you are writing out the contract, work out a payment agreement. Do not be shy about asking the subcontractor how he is paid; they discuss this with every person who hires them.

Do not ever agree to pay the full price up front. You will be left with no leverage if you do this. Instead, the written agreement should normally call for draws. This is to be expected. The best way to ensure that a sub will finish a job is to withhold money. Do not hold back money that you owe him; rather, but do not pay everything up front. By agreeing to certain percentages as you go, you can be sure that the job gets completed.

Here is an example that will show you why this is important. Imagine that you have hired someone to put in your heating system. Halfway through the job, he goes broke. If you have paid him 75 or 100 percent of his total bid, you will not have enough to pay a new HVAC installer to finish the job. You are left without heating and cooling, without the funds to pay for it, and perhaps an inspector coming tomorrow. What will you do?

Here is another example. Don C., an owner-builder, hired an electrician to do all the wiring on his home. The man asked for 60 percent up front and 40 percent at completion, which seemed reasonable — until the county inspector showed up. Some of the wiring did not meet code, but Don had already paid the electrician; his phone calls were not returned. Don paid an extra $2,000 to a different electrician to resolve the issues.

For these reasons, there needs to be some sort of arrangement so that you can safeguard yourself, but at the same time your subcontractor has been paid for his work. Below is a rough guide to the draws you might agree on. Remember that these may vary; different subs may have established a different plan that works for their company, and trends may vary by geographical location as well.

SUBCONTRACTORS' DRAWS

It is important to decide up front with the sub how he will be paid — and then stick to it! Do not pay a sub for work he has not yet completed. Some subs live for their Friday afternoon beer money, and others are hard-working family men who would not dream of skipping out on the job before it is finished. You truly have no way of knowing which one the sub is you have hired, unless he is your brother. Assume that he is the beer money fellow, and do not pay him if you have not agreed to do so. Of course, if you owe a certain percentage because that is your agreement, do go ahead and make the payment on time, without waiting for him to ask for it.

Do not pay for work that has been left incomplete when the sub has told you it is. Always inspect to be sure this is not the case. Sometimes people get in a hurry to receive the paycheck and rush through the job. Make it a policy to always inspect the work before paying. As the subs see you doing this they will pass the word that you will not pay unless the work is done.

Do not pay the entire sum up front. If you pay your sub he will never come back. Agree on a certain percentage up front if he requires it, then fill out a schedule of the work he will perform and the way he will be paid.

Carpenters' Draws

The carpenter's draw is a little different from everyone else's. A carpenter is paid as he does the work. This will take a little communication on your part if you do not have the payment agreement written in a specific way in the contract. The carpenter gets a draw based on the percentage of work he has completed. So if he is about 30 percent of the way done, pay him a 30 percent draw.

Plumbers' and Electricians' Draws

Forty percent upon completion and inspection of any rough-in, and the remaining 60 percent to be paid after the job is complete and you have had the final inspection.

Brick Masons' and Painters' Draws

These subs will require a draw during the process. It is best to inspect the job and compare what they have done so far to what remains to be done. Do not pay ahead. Instead, pay smaller increments if needed so that you stay right in line with what they are doing.

GETTING RESULTS FROM YOUR SUBS

One of the fastest ways to get a sub to complete the job — and maybe even to do a great job — is to tell him that you will pay cash. Cash is a good incentive to get things done. If you are concerned about the IRS, write out a check and then cash it, marking it as his pay. By doing this, you will have a paper trail to show that you had the work done.

When you think about "getting results," remember that results are only what you have gotten in writing. If you are asking for specifications that are not in the contract, it is not likely that you will get it. Technically you did not ask for it. So be sure that the contract words things your way with every specification.

PAPERWORK

It is a fact that many subs hate doing paperwork. You may find that it is difficult to get some to write out bids or to sign documents. This is not something to take personally; it is merely a fact of working with certain trade professionals. The only things that are necessary are the contract and the invoice marked paid in full. If you absolutely must hire someone who works this way, consider writing out the contract yourself, and insist that he present you with an invoice marked paid. These two steps will protect you should problems arise later.

CONTRACTOR CHECKLIST

This is a list of some of the potential contractors that will be needed in the course of construction:

- Architect
- Key Carpenter
- Surveyor
- Soil Treatment Company
- Footing Contractor
- Waterproofing (basement)
- Plumbing
- Well and Septic
- Electrical
- HVAC
- Roofing Contractor
- Insulation Contractor
- Drywall Contractor
- Paint Contractor
- Flooring Contractor
- Countertop Contractor
- Landscape Firm
- Fee Appraiser
- Tile Contractor
- Cleaning Firm
- Brick Mason (to build block foundation)
- Concrete (walls, slab/floors, driveway, walkway)
- Grading and Excavation Contractor
- Finish Carpenter (you may use your main carpenter for this)

CHAPTER 7 CHECKLIST

- Generate a list of potential subcontractors for each necessary craft (see list above)
- Get referrals and recommendations
- Prepare bid requests and get bids
- Check out leading bidders
- Select subcontractors
- Write contracts including payment schedules
- Get all necessary signatures and make extra copies; store copies separately from each other or scan into computer

Renovating a Home

There are many good reasons to renovate a home, and if you have decided that a renovation is the best choice for you should proceed in the same way that you would to build a new home. Create a budget, determine what you need done, and interview contractors.

THE HOME APPRAISAL

When choosing to renovate or remodel, there are a couple of extra steps to take to protect your investment. The first step is to have the home appraised. Explain to the appraiser that you are planning to renovate, and let him or her give you the value of the home when the remodeling is complete.

THE HOME INSPECTION

The other important step to take is to have the home inspected by a professional, who can tell you exactly what condition the home is in before you begin. A comprehensive inspection will give the status of critical issues in your home. For example, is the electrical circuitry safe and polarized correctly? Is moisture getting into the house? Are structural supports in place as they should be? What is the condition of the firebrick on the chimney? Is the chimney properly aligned? Finding out about these things may change the scope of the renovation; it will at least let you know up front about hidden problems before you begin to remodel and you can plan your strategy and your budget. Often when people delve in without an inspection, they uncover problems as they go and have to react to them, rather than being proactive. The few hundred dollars you spend on the inspection will be quickly earned back because of the time you will save on the project.

Note that a home inspection is not like an appraisal. It does not tell you what the house is worth. Instead, the inspection tells you about the condition of the home. It will tell you whether there are components of your home that need major repair work or must be replaced.

At the American Society of Home Inspectors' Web site, you can find a "virtual home inspection" link, which will walk you through a step-by-step inspection and help you understand why inspections are useful. Are the walks safe? Is the plumbing keeping water and gasses inside the pipes and away from the interior of your home? Problems with pipes cannot always be seen, but the rust in the pipe will accumulate and eventually restrict the water flow.

Are breakers, fuses and disconnects as they should be? Is the service enough for the house?

Is the heating as it should be? Are the vents, flues, and Central and through-wall air conditioning safe? Make sure there is an overflow plan so that condensation is carried away from the house. On the interior, study the safety of both the insulation and the ventilation.

These are just a few of the thousands of details an inspector will look at. Plan on the inspection taking as long as two or three hours.

Armed with the inspection and the appraisal, make a decision about the project. If the appraised value after the renovation will make your home the highest priced house on the block, it will be very hard to sell in the future. You may even lose money when you try to sell it because you could end up accepting a price that is more in line with the prices of the neighbors' houses. Remember that your costs may end up being much higher than you anticipate, so the value comparison may be far off from reality.

ENVIRONMENTAL TESTING

If you did not have tests done on your home when you originally purchased it, it may be beneficial to test for the presence of hazardous materials that could jeopardize your family's health. The State Department of Environmental Protection is a good source for finding out about testing in your area.

Radon is a radioactive gas that occurs naturally in the soil. It is the second leading cause of lung cancer in the US; nearly one in every 15 homes has elevated radon levels. It does not matter whether your home is new, old, sealed, or has a basement — the gas rises from the ground to the air and into your home through the foundation, cracks or gaps, or construction joints. If you or your inspector find high levels of radon in the home, it is important to seal foundation cracks and other openings. Installing a vent pipe system is also a common way to remove the radon from under the home and vent it to the outside. If your renovation involves the conversion of an unfinished basement to a living area, test it for radon before you begin. If you find radon, the average cost of repair is only $1,000 to $2,000. Radon can also enter through well water. Sometimes building materials give off radon, although this is rare.

Asbestos, used to insulate homes as well as for flooring, is present in nearly every house that was built between 1920 and 1960. Asbestos can be found in siding, flooring tiles, and pipe insulation. It does not have to be removed, and if it is left alone and in good condition it does not pose a health hazard.

But if asbestos is damaged by your renovation efforts, it must be abated or removed. Asbestos can be covered over with new material if you are careful not to break it and if it is completely covered by the new material. If you must remove and dispose of your asbestos, it is expensive. A licensed professional must do the removal.

Another substance found in pre-1960 homes is lead. You can find lead in two places: the paint and the water system (because lead solder was used on the pipe joints). Lead poisoning is dangerous, especially to young children. In spite of the best efforts of realtors and the government, there are still children becoming ill due to lead-based paint. One child, who was found to have 10 times the allowable amount in his system, was determined to have gotten the paint from a window casement where the family often opened and closed the window. He was not putting the paint in his mouth, but because his face was at the level of the window ledge and the paint was peeling, he still became contaminated by the paint.

There are lead paint test kits on the market that you can perform yourself. You can also hire a licensed professional, who is able to recommend ways to remove lead if it is found. Lead-based paint can be easily covered over with special paint.

If there is lead of over 15 ppb or other contaminants in the water, contact your local health department. They will have information on professional remedies. However, there are steps you can take to prevent poisoning:

✖ Let the tap water run for 15 to 30 seconds before using it.

✖ Do not cook with water from the hot tap or drink it, because hot water dissolves lead faster than cold water.

✖ If you are just building a home, remove the faucet strainers from all the taps and run the water for three to five minutes.

If the copper pipes are joined by lead solder, done since 1986, then the plumber did the work illegally. You can ask that he replace the lead solder, or you can notify the health department.

Determine whether the service line connecting your home to the water line is lead by asking the home inspector or the local building inspector to look at it. However, municipalities are not required to remove lead water lines.

During renovation projects, lead dust can be breathed in; lead can also enter through drinking contaminated water or by eating paint chips.

OTHER POSSIBLE HAZARDS

There are numerous other items that can lead to environmental hazards or "sick" houses. If you are renovating the home you already own or you have purchased one for that purpose, pay the small fee for a professional inspection. Here are some of the possible dangers:

- Aluminum wiring

- Chimney and flue defects

- Drinking water — arsenic

- Asbestos

- Fiberglass

- Past flooding

- Electromagnetic fields

- Oil storage tanks, either above or below ground

- Insulation or ventilation issues

- Sewage or septic backups leaks and clogs
- Structural defects

EXTREME MAKEOVERS

If you are considering a major overhaul, there are plenty of pros and cons on both sides. Remodeling means a lot of hard work, subcontractors that in and out of your house at all hours, and a huge expense. Will you recoup your investment? In most cases, you can expect a return of 60 to 80 percent on your investment, depending on what job you undertake.

On the other hand, if you sell the home there are still expenses: real-estate fees, moving costs, and a few thousand in closing costs. So when does it make sense to remodel?

RETURN ON INVESTMENT FOR SOME PROJECTS

Remodeling Magazine calculates the cost versus value of remodeling projects every year. Although the actual price varies depending on which area of the country you live in, at least you can get an idea of which projects might be valuable for you. Here are some of the more popular renovations:

MID-RANGE PRICE	COST RECOUPED
Attic Bedroom Remodel	76.6 percent
Bathroom Addition	66 percent
Deck Addition	85 percent
Family Room Addition	68 percent
Home Office Remodel	57 percent
Siding Replacement	83 percent
Two-story Addition	73 percent
Window Replacement	80 percent

WHEN FEELINGS INTERFERE

The biggest issue when remodeling a home that you already own is the emotional investment, which is something that cannot be measured. If you live on the same street as many of your relatives or closest friends, for example, you may not care about returns on investment; you simply want your home the way you want it so that you can live there.

Sometimes feelings go the other way — you are sick of living in your house and you want to move on. If this is the case, by all means consider building your dream home from scratch. Remodel the home you live in only enough to make it as marketable as you can; use the rate of return guide to decide which projects you will undertake. If the return brings the value of the house up, it is a good investment. In some cases, the renovation brings a dilapidated house up to the level of the neighborhood. If this is the situation you are in, the renovation is almost a must — unless you are willing to accept a very low price for your property.

Sometimes you love your location, just not your house. There may be traffic flow problems, water problems, or mold. It may simply not fit your ideal. Gutting or tearing down a house may not always be suitable, but there are times when the location of the house or the beauty of the lot supersedes the desire to buy something else. If you are thinking of tearing down the house and starting over, there are a few things to think about first.

Do you have a mortgage on the property? If so, it is actually illegal to tear down the house. One lender said that his institution would allow it, but only if the total balance owed on the loan is less than the value of the land because the land will be all that is left.

How restrictive are the zoning laws in your area? Do you have limitations in your Homeowners' Association covenants? These will limit your remodeling options, and may cause you to rethink doing a major remodel. If you do decide to do a major renovation, remember that many municipalities charge a lower property tax rate on remodels as opposed to new construction. So leaving part of the old home standing can save money at tax time.

Besides the restrictions caused by community covenants, there are the neighbors to consider. In most cases the planning department requires that the community be notified of major renovations. The neighbors then have a voice in the decision. Sometimes there is even a design review board that must approve your plans. Be sure that your neighborhood will tolerate your plans; whether you are going to underbuild, overbuild, or fit perfectly into the neighborhood design, their tolerance matters.

How long do you plan to be in the home? If you are going to be there for a long period, you are improving your own lifestyle by doing a major remodel. If you are not staying in the home for a long time, or you are unsure of the length, then it will be more important to assess the impact of the renovations against the resale value.

Can you find a reasonably priced house with all the amenities you want? If the supply of homes in your area is limited, and you already enjoy your home, a remodel is the perfect choice. You can make the house exactly what you want it to be.

How does the house compare to others in the neighborhood? If it is in need of siding, has only one bathroom while all the neighbors have three, needs updating, and has no distinctive features that will make it stand out from the crowd, then it needs a remodel whether you are staying there or not.

If you do decide to demolish an existing home, do not forget to add the demolition and removal costs to the expense of your project. If the property contains hazardous waste, which is common in old houses, it can cost even more. Get an estimate for both the demolition and the removal of all materials before making a firm decision.

REMODELING CHECKLIST

1. Analyze your Budget

2. Analyze the Costs

   Be sure that you include:

   - Permit fees
   - Construction materials
   - Labor costs (you will need to know the total hours)
   - Tools and equipment
   - Safety equipment
   - Materials
   - Incidental repairs (which are inevitable when you delve into an existing structure)
   - Cleanup (including rental bins, trash removal or dumping)
   - Decorative items
   - Other expenses

3. Consider the Return on Investment

   Market valuation — If the home prices in your area are increasing, the renovation may add good value to your home. If the market values are decreasing, as many are at this time, the renovation may not add value to your home.

Home comparison — How will the renovation affect the look of your home as compared to others in the neighborhood? If your project does not fit, in some instances the value is decreased.

4. It is impossible to predict whether the market is moving up or down in a given area, but there are some factors that can give a particular impact. One is:

   Construction — New highways or other construction may impact the value of your home later. In the 1980s, I built a home with the understanding that the farm it backed up to was in the family for generations and will never be sold. Just after my home was complete and we closed on the loan, the same agent who uttered those words began developing the property. Suddenly, my idyllic setting was full of construction noise, pollution, and a row of cookie-cutter houses, which far undermined the value of my own.

5. Check with the local government to see if there are zoning changes afoot: **www.statelocalgov.net.**

6. Study your boundaries. An open field may become a new development, a commercial zone may encroach on your privacy, or a road that is opened to a bigger thoroughfare may become heavily traveled.

7. How is crime in your neighborhood? Are the crime rates up?

8. How much equity do you have in your home? If your remodeling project is more than 30 percent of the value, it may be better spend on a new home.

9. Conduct Appraisals and Inspections

10. Reevaluate Cost (if necessary)

11. Develop the Plan

    If you have taken all the steps and decided to go forward with the plan, create the specs and check with building codes for the details about meeting code.

12. Develop Financing

    Use the equity in your home to obtain a home equity loan or line of credit. Consider the loan to value (see Chapter 4) and the qualifying debt to income ratio if the loan is a significant amount. If the size of the loan versus your equity makes this kind of loan impossible, consider alternative financing mentioned in Chapter 4.

13. Manage the project just as you would a new construction

CHAPTER 8 CHECKLIST

- Have the home appraised and inspected
- Have the home environmentally tested
- Review the "Other Possible Hazards" list and determine which need to be checked
- Review your potential investment versus resale payback versus other alternatives
- Develop the plan
- Develop the budget (see Remodeling Checklist)
- Finance the renovation
- Manage the renovation as you would new construction

Comply with Building Code, Inspection, & Permit Requirements

The is a Uniform Building Code (UBC) in effect, and most building departments rely on it at least somewhat. Building Codes exist in every jurisdiction in the United States. Some are stricter than others; some have the same rules but interpret them differently. Some must create special rules because of special conditions that are indigenous to an area, like snow loads, earthquakes, and hurricanes. This can make it hard for you as a new builder to attempt to follow the Building Codes. The best thing to do is get to know the people in the local code enforcement office who will be handling your project.

DESIGN REVIEW: UNDERSTANDING WHAT IS ALLOWED

If you are building in a neighborhood with many homes that all look alike, you can assume the city, county, or neighborhood commission has developed specific guidelines governing what can be built there. If your plan is to build in such a community, expect strict guidelines and much delay. You can begin by getting a copy of the local Building Code from the local government for neighborhood guidelines and formal guidelines, called CC&Rs, for the subdivision.

Before spending thousands of dollars for formal plans, take the extra step to get the preliminary design approved. Make sure that everything you have in mind will pass the approval, and change the design to fit if you find that it will not pass. This will save a lot of headaches later on — and possibly save you from tearing down part of your house!

BUILDING PERMITS

Homeowners are able to get all the permits needed in most jurisdictions in the United States. But it is not necessary for you to get all of them; many of the licensed tradesmen you hire as subs will provide their own permits.

To get a building permit, you will simply supply two sets of prints and specs to the local code enforcement office. The authorities will examine the plans and either approve or reject your project.

After the plans are approved, one set of prints will be given back to you. The other will be kept in the code enforcement office. You must keep the approved set of plans at the job site at all times. It is stamped and marked as approved; the inspector will need access to it when he comes on site.

In addition to the plan approval, the code enforcement office will issue you a permit. This has to be displayed prominently at the job site. Wrap it in plastic to protect it and mount it high on a tree or pole so that it remains safe and unscathed from knocks, bumps, or vandalism.

Some of the other permits may be kept in your files; others must be posted. The plumbing, heating, and electrical permits, acquired by your subs, must not only be posted, but must also be visible from the road.

PERMIT CHECKLIST

The required permits vary from one area to another. But the permits that will be necessary for your project are:

- ✖ Building Permit
- ✖ Plumbing Permit
- ✖ Heating Permit
- ✖ Electrical Permit
- ✖ Septic Permit
- ✖ Driveway Permit
- ✖ Inspections

Inspections were covered in Chapter 6, but they are such a large issue for so many owner-builders that this bears repeating. Getting plans approved and proper inspections done may go smoothly, but they are often a nightmare. Assuming your regular job is not in the realm of architecture, engineering, or contracting, you do not know the process or the people who are involved.

The easiest way to offset this is to learn all you can about the steps required, familiarize yourself with the rules and regulations of your local government, and follow the steps that they require (called a plan check). Be prepared to provide them with all your plans and whatever else they ask for. It is helpful if the contractors you are using are familiar with these departments, so that they know to whom you should submit to and how to go about making the process quick and smooth.

One inspection that is extremely important is the one that precedes the driveway permit. If this is required in your jurisdiction, an inspector will look at the spot you have chosen for your driveway to enter the main road. He will determine whether the location is safe. If it is, he will tell you what size culvert pipe you need and issue you a permit. In some cases the location is not the best due to the lay of the road, so you may be asked to move the driveway.

Failing to have any of the necessary inspections can result in much extra work and expense. For example, if you were asked to move the driveway, imagine what it would cost to remove the concrete and replace it. Or if you forget your plumbing inspection, you could be asked to remove walls and fixtures to allow the inspection to take place.

BUILDING INSPECTORS

The county inspectors will review the work done and certify it. This has to occur at certain steps of remodeling or construction. For example, temporary electrical service will be inspected to ensure that it is properly grounded. The slab will be inspected prior to pouring concrete to be sure that it is insulated correctly. Footings will be inspected to be sure that they are resting on solid (load-bearing) ground. It is important for you to be present at every scheduled inspection. The inspectors from the county or municipality are not inspecting the quality of the work; they are making sure it meets code. The final systems like insulation, electrical, plumbing, HVAC, and others must be checked to ensure that they:

✖ Work

✖ Meet code

✖ Are safe

What kinds of codes might be enforced? Some areas restrict the size of homes to one or two stories. Some require that driveways be asphalt or cement. Some restrict lots to a certain size or kind of house. Homeowners' Associations might state that you have to use stucco or cannot use stucco. They might require a certain size house or a certain size window. These requirements are all found in the CC&Rs that are a part of the land.

After these final inspections, a certificate of occupancy will be issued.

BANK INSPECTORS

The inspectors who represent the lender(s) arrive at the end of each phase; their purpose is to see that the money borrowed is being used as it was intended. These are not building inspectors, so they are not looking into quality or code. They are only taking photos and keeping records.

Normally, subcontractors schedule their own separate inspections; these are also inspections that you should attend. How do you make sure that the work done by subcontractors passes inspection, meets code, and also is exactly what you requested? The simplest way is to hire your own, independent inspector. This person can "pass" the work prior to you making any payments to the sub. You are completely within your right to withhold payment until the subcontractor has delivered all the services that are listed in your contract. By hiring a professional inspector, you are both protecting your investment and creating a paper trail, should the need for one arise. Obviously, this stresses the importance of a detailed, comprehensive construction contract, and also the need to thoroughly discuss all project plans with subcontractors.

WHEN CONSTRUCTION FAILS INSPECTION

If any step of your construction "fails" its municipal inspection, it means that step did not meet code. You simply make the corrections or adjustments and request that a follow-up inspection be conducted. There will be an additional re-inspection fee. If your independent inspector finds fault with any part of the construction, he must explain the reasons, which you can use to discuss the need for corrections with your subcontractor. Independent inspectors do not "fail" items, they simply point out the physical conditions and show you what parts may need replacing or repair.

Remember that inspections are a positive, not a negative. There might be as many as 15 inspections during the building process. They may seem like a hassle to schedule, and it often feels as if the inspection process holds up the project. But they ensure that the work is done properly and they also protect your investment.

CHAPTER 9 CHECKLIST

- Get copies of local Building Codes and study in light of your plans
- Review the Permit Checklist; ensure you have all needed permits
- Understand what inspections will need to be made and what the inspector will be looking for; remember, they are your friends!

10

Building Materials & Foundations

WHERE TO PURCHASE BUILDING MATERIALS

To purchase materials, you will first need a list of all the supplies you might need. This is called a take-off list. In this chapter there is a master list to help get you started. Supplier prices, as you are about to find out through personal experience, vary wildly. You will need to get several quotes on each kind of supply to find the best prices.

Where will you get the quotes? If you have already selected subs to do the work, ask them where you should purchase your materials. They often know who has the best pricing or who will offer a discount. Do not be tempted to let the sub buy the materials for you, in case there is a markup. Instead, go to the supply house and negotiate your own price.

When you purchase plans, there is often a materials list supplied for an extra fee. This list is something that can be used as a guideline, but it is not infallible. Building Codes and geographical climates create differences in the materials used in various parts of the country. Even the wall studs vary; some areas may use 2x4s and others use 2x6s — so if your list calls for one instead of the other, you will end up with the wrong supplies.

Should you let the supplier create your list? As a test, try this. Choose several suppliers and ask each of them to create the take-off based on your plans. Most suppliers are willing to do this for free to entice you to do business with them. In the meantime, create your own list of supplies based on the list in this book. When you gather all the lists, sit down and compare them. You will find variations; in some cases, they will have forgotten to list certain items altogether. In others, the prices or amounts figured will vary. There are always several ways to do a job, of course, but it is better to create the list yourself.

Building materials may come from:

- Lumberyards
- Builder's Supply Stores
- Do-It-Yourself Superstores
- Builder's Closeout Stores
- Item-specific Warehouses
- The Internet
- eBay

HOW TO SAVE ON BUILDING MATERIALS

Shopping for building materials is just like shopping for any other consumer product. You can buy in bulk and save; you can buy at the superstore and save some; or you can buy at a specialty store and pay premium prices. The prices you pay will also depend on your purpose. For example, a professional builder considers time a more important factor than price, so as long as the price he gets is fairly competitive, he is not going to spend a great deal of time shopping around for a deal. He could make a little more profit by comparison shopping, but if he is earning a $30,000 profit building your $275,000 house, then he would prefer to move forward by building rather than shopping. Plus, the builder may be earning as much as 30 percent on the items you select for your home.

But as an owner-builder, you do have time to look for bargains. You can shop and save — the more time you are willing to invest in the search, the more you will save. Some of the contractors you speak to will try to tell you that they get a builder's discount that you cannot match, but this simply is not true. Individuals can find terrific savings on any number of materials by simply looking for deals. Some of the savings owner-builders have mentioned include:

Staircase parts — $14.99 at the local builder's supply, $12.99 at Home Depot, $$3.75 at the builder's surplus store 20 miles away. The homeowner needed 98 of these; the savings totaled $1,101.

Tile — One homeowner located a tile supplier four hours away. Since they were installing 1800 square feet of tile flooring, he and his wife chose to make the drive. By choosing closeout tile, they paid $0.99 per square foot instead of $11.88 — for a total savings of $19,602! This was well worth the eight-hour round trip drive to the supply store.

Windows — High quality windows are costly, yet it is possible to find terrific deals. I myself ran across a couple whose renter had ordered $36,000 worth of custom windows — then skipped town, leaving the owners to pay the bill. They sold the windows for $3,600 just to get them out of the way. A self-employed installer altered the design of the 15 windows in the house and put them in for $1,800. An added bonus for the homeowner was that switching to bigger windows let in a lot more sunlight.

Appliances — Do not put off shopping for appliances until the job is nearly complete! It is possible to find closeouts, end-of-year models, and other deals. My own latest washing machine was purchased at Sears for $300 less than the listed price. Its model number was the same but the SKU number differed from the one that was on "sale" for $150 off. There was no difference in the product, only in the one number, which was printed in tiny writing on the display at the store. They had two of these still in the warehouse; I presume it was simply last year's version of the same washer. No employee could explain why it was different. I had to ask them specifically to type in that special SKU number to get the right price. After I asked, they even threw in free delivery. Keep your eyes open and check those part numbers!

Siding — Siding is one of the products that has the biggest variation in price (windows are another). Siding quotes vary as much as 30 percent from one supplier to another. Subcontractors are especially helpful in telling you where to shop for siding. One homeowner purchased through an individual who sold siding on the side and saved 50 percent of the previous lowest quote he had received.

The point is that there are savings on every single item in your home; you simply have to be willing to shop and use your head. Watch the newspapers for sales; check out all the local closeout stores; and do not be afraid to talk about your project and ask questions. One homeowner talked with a builder's supply store manager, who did not offer the best prices. But the manager remembered him, so when some windows were returned to the store, he phoned the homeowner — who got them at 25 percent off the retail price.

If there is an area fairly close by where the wage is likely lower, shop there. Even the chain stores have different pricing for different geographical locations. Driving an hour or two to save 10 cents per part does not make sense, but when you are purchasing flooring, pipes, windows, or sheetrock for an entire house, you could be saving in the thousands.

SHOULD YOU PURCHASE ALL THE MATERIALS FROM ONE SUPPLIER?

This is a question to seriously consider. It is not the best way to save money. Often you find that one supplier has great prices on lumber, but not on siding. Or they have good prices on doors and windows, but they are the highest quote for shingles and trusses. However, there are several advantages to using just one supplier. You will get better service if they are supplying the entire job. You can ask for a price lock, so that the price quotes you get up front are good for the duration of the project. And, with only one supplier, the paperwork is easier; you only have one set of invoices, one place to make returns or discuss shipment errors, etc.

Whether you decide to use one supplier or several, always ask questions. Ask whether that is the best price; ask whether you can get an extra 10 percent off if you purchase all the parts at this location. Ask for free delivery. Ask to see the discontinued items, damaged stock, and overstock.

It is always possible to save by purchasing the materials separate from the labor. If subcontractors are giving you bids with the materials included, they are using a formula to compute the cost of materials. By looking at the breakdown and comparison shopping, you can also beat that price — then find a subcontractor who is willing to do labor only. This saves money on the product you are purchasing, and allows you to control the quality of the materials; you will know exactly what you are getting. You will have the warranty information at hand, and you will be able to get a refund if you have overestimated the amount of material that you need.

Furthermore, you may even save on labor; you will see that you can install some of the components by yourself, and you can negotiate with subs for labor-only until you find a good price.

TAKE OFF LIST INFORMATION

Plan Name / Date	Sample	
Plan Description	Residence	
Customer / Phone	Private Customer	11.222.35
Contractor / Phone	D&D	800-555-1212
Supplier / Phone	Sample Supply Store	111-555-3333
Deliver To	8766 Pine Tree Ln. Mytown, Mystate 55555	

Foundation/Drain Materials

21 pcs.	AB0812	Anchor Bolts for hold down
77 pcs.	1/2x8"	Regular J-bolts for mud sills
18 pcs.	16x8x8 "	Vents for crawlspace
9 sheets	2" R10.5	Rigid board 4x8 for slab insulation
250 lin. ft. 4"	4"	ADS solid pipe for downspout drain
235 lin. ft.	4"	ADS perforated pipe for perimeter drain
3 pcs.	4"	Connectors for drain pipe
8 pcs.	4"	Elbows for drain pipe
16 pcs.	4"	Tees for drain pipe
4 pcs.	4"	Ys for drain pipe

Sill Plates

236 lin. ft.	2x8"	Treated Sill Plate (check code; 2x8 gives more stability than 2x6) for mud sills
5 rolls	6"x50'	Sill Sealer for mud sills
132 pcs.	1/2"	Nuts & Washers for anchor bolts
Basement Floors		
1 pcs.	9'	Steel I-beam for basement
1 pcs.	8'	Steel post (if needed) for basement

Walls — Basement

172 lin. ft.	2x6"	SPF for plates
102 pcs.	2x6:	SPF, 92-5/8" for studs

TAKE OFF LIST INFORMATION

64 lin. ft.	2x6"	SPF for blocking, bracing, etc.
86 lin. ft.	2x4"	SPF for plates
32 pcs.	2x4"	SPF, 92-5/8" for studs
44 lin. ft.	2x4"	SPF for blocking, bracing, etc.
20 sheets	3/4"	Birch Plywood, 4x8 , for sheet siding
9 rolls	1000 sq. ft.	Typar for building paper
2 roll	2"	Tape for building paper
54 tubes	10 oz.	Silicone Latex (white) for caulking
4 pcs.	HDA	for holddown
16 pcs.	5/8"x4-1/2"	Stud (hex) Bolt Set for HDA
4 pcs.	5/8x24"	Steel Threaded Rods for HDA
8 pcs.	5/8"	Nuts and Washers for HDA
18 pcs.	HD5A	Holddown
40 pcs.	3/4"x4-1/2"	Stud (hex) bolt set for HD5A
20 pcs..	5/8"	Nuts and Washers for HD5a
20 pcs.	5/8"	Nuts for threaded rod
14 pcs.	MST	Metal Strap Tie for walls
1 roll	CS16	Metal Strapping for walls
13000 pcs.	16d	V.C. Sinker Hitachi for framing nails
40 lbs.	1-1/2"	Teko Nails, galvanized, for connector nails
25000 pcs.	8d	Galvanized Nails for sheathing nails

Floors — 1st Level

2180 lin. ft.	11-7/8"	TJI for joists
386 lin. ft.	1-1/4"x11-7/8"	Timberstrand for rims
118 lin. ft.	11-7/8"	TJI for blocking
80 sheets	3/4"	Birch Plywood for subflooring
25 tubes	PL400	Construction Adhesive for subflooring
28 pcs.	JS210	Joist Hangers for 10-1/2" I joist
2100 sq. ft.	4 mil	Black Polyethylene for crawlspace
5000 pcs.	8d	Gun Nails for subfloor nails
1 pcs.	8x15", 12'	Beam for great room
1 pcs.	4-1/ x12", 8'	Beam for kitchen opening

TAKE OFF LIST INFORMATION

Walls — 1st Level

1260 lin. ft.	2x6"	SPF for plates
264 pcs.	2x6"	SPF, 92-5/8" for studs
120 pcs.	2x6"	SPF, 10' for studs
240 lin. ft.	2x6"	SPF for blocking, bracing, etc.
280 lin. ft.	2x4"	SPF for plates
260 pcs.	2x4"	SPF, 104-5/8" for studs
432 lin. ft.	2x4"	SPF for blocking, bracing, etc.
156 lin. ft.	2x6"	SPF for headers
64 lin. ft.	2x8"	SPF for headers
98 lin. ft.	2x10"	SPF for headers
24 lin. ft.	2x12"	SPF for headers
32 lin. ft.	6x12"	SPF for headers
124 sheets	7/16"	OSB 4x8 Panels for sheathing

Floors — 2nd Level

1660 lin. ft.	9-1/2"	TJI for joists
285 lin. ft.	1-1/4"x9-1/2"	Timberstrand for rims
68 lin. ft.	9-1/2"	TJI for blocking
38 sheets	3/4"	Cedar Plywood for subflooring
24 tubes	29	PL 200 Adhesive for subflooring
26 pcs.	JS08	Joist Hangers for 8" I joist

Walls — 2nd Level

880 lin. ft.	2x6"	SPF for plates
264 pcs.	2x6"	SPF, 92-5/8" for studs
168 lin. ft.	2x6"	SPF for blocking, bracing, etc.
464 lin. ft.	2x4"	SPF for plates
164 pcs.	2x4"	SPF, 92-5/8" for studs
260 lin. ft.	2x4"	SPF for blocking, bracing, etc.
90 lin. ft.	2x6"	SPF for headers
128 lin. ft.	2x8"	SPF for headers
88 sheets	7/16"	OSB 4 x 8 plywood for sheathing

TAKE OFF LIST INFORMATION

Floors — 3rd Level / Other

48 pcs.	2x10"	DFIR (2), 16' for joists
132 lin. ft.	2x10"	DFIR (2) for rims
22 lin. ft.	2x10"	DFIR (2) for blocking
1220 lin. ft.	2x6"	T&G Car Decking for subflooring
12 pcs.	LU210	Joist Hangers

Walls — 3rd Level / Other

466 lin. ft.	2x6"	T&G Car Decking for subflooring
142 pcs.	2x6"	SPF, 92-5/8" for studs
84 lin. ft.	2x6"	SPF for blocking/bracing
80 lin. ft.	2x4"	SPF for plates
24 pcs.	2x4"	SPF, 92-5/8" for studs
36 lin. ft.	2x4"	SPF for headers
48 lin. ft.	2x6"	SPF for headers
24 lin. ft.	2x8"	SPF for headers
44 sheets	7/16"	OSB 4 x 8 Plywood for sheathing

Roof Framing Materials

1 each		Truss Package
6 pcs.	2x6"	DFIR (2), 8' for rafters
8 pcs.	2x8"	DFIR (2), 12' for rafters
8 lin. ft.	2x8"	DFIR (2) for ridges
12 lin. ft.	2x10"	DFIR (2) for ridges
2 pcs.	2x6"	DFIR (2), 12' for valley ledgers
2 pcs.	2x8"	DFIR (2), 16' for valley ledgers
75 sheets	1/2"	CDX Plywood for roof sheathing
33 sheets	7/16"	OSB, LP Primed for eave/overhang
4 sheets	7/16"	OSB 4x8 for gable sheathing
270 lin. ft.	2x6"	Cedar for fascia
58 lin. ft.	2x8"	Cedar for rake
60 lin. ft.	1x3"	Cedar for rake trim
124 pcs.	H2.5	Hurricane Ties for rafters / trusses
378 pcs.	H-Clips	Plywood Clips for sheathing
2 pcs.	12x18"	Gable Vents

TAKE OFF LIST INFORMATION

Interior Stairs

9 pcs.	2x12"	DFIR (2), 18' for stringers
28 pcs.	1-1/8x12"	OSB B.N., 3' for treads
10 pcs.	1-1/8x12"	OSB B.N., 12' for treads
20 pcs.	1x8"	Utility, 10' for risers
12 pcs.	1x8"	Utility, 12' for risers
18 pcs.	LU28	Joist Hangers for stringers

Window Schedule

4 pcs.	5'0x5'0	Vinyl, Anderson, white, top awn
2 pcs.	4'0x5'0	Vinyl, Anderson, white, top awn
1 pcs.	3'0x5'0	Vinyl, Anderson, white, top awn
2 pcs.	3'0x2'0	Vinyl, Anderson, white awning
2 pcs.	2'0x4'0	Vinyl, Anderson, white awning
2 pcs.	2'0x4'0	Vinyl, Anderson, white, fixed
4 pcs.	4'0x5'0	Vinyl, Anderson, white, top awn
3 pcs.	3'0x5'0	Vinyl, Anderson, white, s.h.
3 pcs.	3'0x5'0	Vinyl, Anderson, white, s.h.
1 pcs.	3'0x2'0	Vinyl, Anderson, white, awning

Door Schedule

2 pcs.	3'0x6'8	Model 8860 dual core door
1 pcs.	6'0x6'8	Entryway
7 pcs.	2'6x6'8	6-panel, hollow core, painted masonite indoor
2 pcs.	3'0x6'8	6-panel, h.c., painted masonite indoor
1 pcs.	3'0x6'8	6-panel, s.c., painted masonite
5 pcs.	2'6x6'8	6-panel, h.c., painted masonite
1 pcs.	2'6x6'8	6-panel, h.c., painted masonite
2 pcs.	4'0x6'8	6-panel, h.c., painted masonite, bi-fold
2 pcs.	6'0x6'8	6-panel, h.c., p.mas, bi-fold
1 pcs.	16'0x7'0	Metal

Siding / Exterior Trim

7020 lin. ft.	8-1/4"	Cedar for siding

TAKE OFF LIST INFORMATION

1.2 squares	#275	Cedar Singles for siding
9000 pcs.	8d	Gun Nails for siding nails
484 lin. ft.	5/4x4"	Cedar for outside corner boards
92 lin. ft.	2x2"	Cedar for inside corner boards
326 lin. ft.	1x3"	Cedar for vertical battens
276 lin. ft.	2x6"	Cedar for frieze boards
290 lin. ft.	2x10"	Cedar for belly band
29 pcs.	1-1/2"	Drip Edge, 10' for belly flashing
364 lin. ft.	5/4x4"	Cedar for casing, windows
110 lin. ft.	5/4x4"	Cedar for apron/casing, windows
110 lin. ft.	1/2x2"	Cedar for parting bead, windows
115 lin. ft.	2x3"	Cedar for sill, windows
133 lin. ft.	3/4x2-1/4"	Cedar for crown mould, windows
60 lin. ft.	5/4x4"	Cedar for casing, doors
16 lin. ft.	1/2x2"	Cedar for parting bead, doors
19 lin. ft.	3/4x2-1/4"	Cedar for crown mould, doors
15 lin. ft.	2x10"	Cedar for side jamb, garage door
18 lin. ft.	5/4x4"	Cedar for head jamb, garage door
35 lin. ft.	5/4x4"	Cedar for casing, garage door
15 pcs.	1-1/4"	Shake Flashing, 10' for door/window flashing
3000 pcs.	16d	Galv. Casing Nails for gun nails for ext. trim

Decks / Exterior Stairs (only pressure treated)

4 pcs.	2x8"	Lumber, 12' for ledgers/rims/blocking
4 pcs.	2x8"	Lumber, 18' for ledgers/rims/blocking
12 pcs.	2x8"	Lumber, 8' for joists
17 pcs.	2x8"	Lumber, 12' for joists
48 lin. ft.	4x4"	Post for rail posts
82 lin. ft.	1x8"	Cedar for trim boards
784 lin. ft.	5/4x4"	Cedar for decking
152 pcs.	2x2"	Cedar, 3' for pickets
80 lin. ft.	2x6"	Cedar for rail

TAKE OFF LIST INFORMATION		
80 lin. ft.	2x4"	Cedar for rail
160 lin. ft.	1x3"	Cedar for rail
40 pcs.	1-1/2"x 3"	Deck Flashing, 10' for deck to wall flashing
42 pcs.	LU28	Joist Hanger for deck joists
300 pcs.	16d	Galvanized Nails for gun nails for deck
2000 pcs.	3-1/2"	Galvanized Screws for deck screws
3 pcs.	2x12"	Lumber, 12' for stringers
6 pcs.	2x6"	Lumber, 10' for treads
3 pcs.	LU28	Joist Hangers for stringers
18 pcs.	TA10	Staircase Angles for treads
140 pcs.	1/4"x1-1/2"	Galvanized Lag Screw for angles
6 pcs.	12x12x8"	Conc. with saddle blocks for rear deck
Interior Trim		
880 lin. ft.	1/2x2-1/3"	Pine for jamb, windows
364 lin. ft.	1/2x2-1/4"	Pine for casing, windows
110 lin. ft.	1/2x2-1/4"	Pine for apron/casing, windows
115 lin. ft.	1x6"	Pine for still/stool, windows
68 lin. ft.	3/4x5-1/4"	Pine for crown molding
68 lin. ft.	1/2x3"	Pine for chair rail
961 lin. ft.	1/2x2-1/4"	Pine for baseboards

SITING THE HOUSE ON THE LOT

If you are like me, you thought the long and arduous task of searching for a lot was the time-consuming part — then when it is time to site the house, you take even longer to decide! That is because it is one of your first real decisions, plus so much rides on the orientation of the house. Will you take advantage of the sun or the shade? Will the house look awkward? Will you regret it after it is complete?

The site cannot be changed once construction begins, so it is ideal to work with the site itself to create a great-looking, functional, cost-effective plan. To do this, there are a few things to consider.

THE NEIGHBORHOOD

Where are the houses sited that are around it? Setting your home further back, further forward, or at an angle compared to the others may make it look undesirable. If the homes are fairly close together, be sure that yours blends with the neighborhood.

This goes for functional elements as well: if the neighbors do not have long porches or front-end garages, you might want to think twice about being the sole non-conformer. Your home does not have to be a cookie-cutter copy of the others, but it will look best if it fits in with its surroundings.

THE VIEW

As the occupant of the home, the view is the most important issue about your house. A view at the back will almost demand that you turn your plan around, with the front, as it is designated on your house plans, facing the back of the lot. Be sure if you do this that the garage is still located closest to the street. The importance of the view will mainly affect the family room and master bedroom, and to a lesser extent, the kitchen and living room. Consider the foliage, mountains, trees, rocks, and bodies of water when deciding how to place the home.

If you are not going to turn the plan around to face the view, you will want to re-design the windows at the very least, so that the biggest windows are on the view side. This is especially a factor if you are keeping the total square footage of windows to a minimum because of any heating and/or cooling concerns.

THE SUN AND THE WEATHER

Depending on the climate of your geographical area, you may want to put the majority of your windows facing southward, where the sun can warm your home in the winter. However, if you live in an area that has mild winters and excessively hot summers, you may be more interested in cooling; you will want the glass to face toward the north, so that the sun cannot beat down so relentlessly in the summertime. This may also offset another issue: storms. For houses in hot locations, the winds that come with summer storms can cool the house.

Normally, storms move in the same direction fairly often in a given locale. That means one side of a house is exposed to more severe weather than the others. The direction varies, but most often the north and west of the house are the bad-weather sides. Hot-weather homeowners will appreciate the north-facing windows. In cooler climes, you can take this into consideration by putting fewer windows on the harsh weather side of the house. You might even locate the garage on the north or west side.

THE SLOPE

Few building lots are flat. Most of the time, a lot will have at least a gradual slope. This will affect the way that people look at the house, how high or low your first floor will be, and the drainage of the lot. It is important to carefully consider all these issues before siting the home.

On a gradual slope downward from the house to the street, the house may seem bigger than it is. You can also be assured that the water will drain, although maybe not in the right direction. Most homeowners want the water to pour toward the front because storm drains are located in the street. This carries the water to the public street and off your private property. But if your water flows onto a neighbor's property, you could end up flooding their property and possibly even being party to a lawsuit. The way the water moves will become crucial once the house is built, so pay careful attention in the beginning.

On a lot that slants downward from the street to the home, the view may focus the visitor on your roof line — and also cause the home to seem smaller than it is. This is not the best position to be in if you should ever choose to sell the property. Plus, a low-lying home can be a victim of runoff, and may require an expensive drainage system to stay dry. If your house is on a slope, look carefully and ensure that water moves around both sides of the proposed home site and down the slope. And if there is an area that seems blocked, either make plans to excavate or include drainage plans in your design.

Sometimes, if a site is particularly steep, a contractor might tell you that the easiest way to build will be to cut and fill. This is when the builder cuts into a slope with a bulldozer to create a spot about half the size of the house, then moves the dirt from that area to level out the rest of the house. The advantage to this is that you will save quite a bit on excavation costs; you will only have to cut into half the dirt, and you will not have to pay and haul it away.

Cut and fill should be approached with extreme caution though — over the long run, the house may not stay stable. Sometimes the filled side will settle, leaving the other side firmly in place. This can actually crack the house down the center. A better idea is to put the entire home either on the fill or on the cut.

Sometimes when lots are sloped, the ground can be terraced or a retaining wall can be constructed to create a building pad for the home. These may be made of steel and concrete, boulders, logs, or railroad ties. They may be placed above or below the building pad. Remember: a retaining wall has to extend at least two feet into the ground for every foot of height above the ground. So a three-foot high retaining wall must be six feet deep. Otherwise, the land — and the house — can slip.

READYING THE SITE FOR CONSTRUCTION

Before beginning any work, hire a surveyor. Even if you recently purchased the lot, it is important to designate the lot lines by marking out all the borders. There are permanent markers that have been installed by the government that correspond to the county plat map. Using the site map, the surveyor, often working in tandem with the contractor, marks out the boundaries, setbacks, and underground utilities. Then they mark the corners of the house and the garage, driveways, walkways, and where utilities will run.

If you have hired a surveyor and a contractor to mark the foundation, go back over all the property yourself; make sure the batter boards or markings have straight lines, square angles, and clear markings. Be sure that the lot corners and setbacks are marked — and measure them yourself so that you know they are correct.

Now is the time to place clear "Danger" signs around the property, and plan where materials and trash will be stored — preferably away from the building site.

Be sure that the surveyor flags the underground utility lines, and make sure they stay marked throughout the building process. Before beginning any work, call 800-258-0808 to ask the utility company to mark where the utility lines run. If you do not call them, and you end up causing damage to a line during excavation, your fine could be as high as $100,000 plus damages. If any of the utilities are in the way of driveways, landscaping, or other items, they can be relocated.

Getting the site work done includes roughing in a driveway, cutting down trees, and possibly installing the well and septic system or digging trenches for the utility hookup. Be sure that you get enough trees cut down to make room for the new home, plus the large trucks, which will have to maneuver around the foundation once it is in. While this is going on, clear a path for the utility trucks to bring in the electrical service. You will save a little money by making sure you get enough trees removed in the beginning. Otherwise you will have to have someone come back and cut more — and pay more for the privilege.

TREE REMOVAL

If you have a lot of trees or if your trees have a large diameter, consider trading the wood for the work. This way you can get the lot cleared for free. Some individuals will remove them to use for firewood. Some logging companies will purchase the timber from you, if it is the right kind and if you have enough of it. However, many of them are scheduled far in advance. I was given a wait time of three years. So check into this carefully before you make it a firm plan. Also make sure that anyone who comes on your property to do any type of labor has insurance coverage.

If you plan to burn firewood yourself, talk to the contractor who cuts the wood about having someone chop and stack it on a corner of the lot. If you do not have room to store it all, sell it or give some of it away. The wood is yours to keep if no other arrangements have been made.

Once you have the trees cut down, there will be stumps to remove. These are large, often as tall as four feet above the ground. Do not be tempted to leave them; they will decompose, leaving ugly sinkholes. They must be hauled away and ground up — and sometimes it takes a hefty piece of equipment to remove them. Do not assume that the person who cuts the trees will remove the stumps; you may have to hire another subcontractor to do this.

It is expensive to remove trees, and it is also costly to transplant new trees. Trees take decades to mature; so if you have mature trees already on the property, it makes sense to utilize them, even if it means redesigning the house or locating it differently on the lot than you initially planned. This is called taking advantage of natural elements.

To protect trees that are on the property which you intend to keep, place hay bales around them to be sure that they are not damaged by heavy equipment. Using bright orange plastic fence from the hardware store will help drivers avoid hitting them, and will help protect the trees from flying debris. Alternatively tie red ribbons around the trees to let workers know not to cut them, and to help them see the trees and avoid hitting them.

Besides tree removal, a bulldozer must come in and clear the shrubs and brush. This is a fairly low-cost activity. After the clearing, it is time for grading; smaller grading projects may even be done by hand, although tractors and bulldozers more often do this job.

GRADING

Grading is the way that dirt is redistributed from one area to another. Scraping the dirt down creates level ground. Leftover topsoil may be held on the lot until it is needed at the end of the project for the landscaping of the property. There will be finish grading at the end of the project, which will smooth the yard as well as ensure that water moves away from the house in a desirable way.

There is not a lot to watch for in grading work, except to be sure that the grading extends three feet or more beyond the edge of the house and garage areas. If the grader discovers lots of boulders, you may have to perform blasting to remove them. Blasting is expensive, and even dangerous, but sometimes it is the only way to remove these rocks.

Occasionally, the rocks are so large that nothing can be done. Several years ago I hired a contractor to bulldoze the back yard. I asked him to dig up or cover over the rocks. There were two of them, each about four to five feet long and three feet wide. They were not a big deal, but they interfered with mowing the yard. He called me outside a few hours later; he had dug alongside one rock and exposed it. Now it was about twenty feet long, and the trench exposed it to a depth of about five feet. The contractor assured me there was a lot more to it. The boulder was much too close to the house for blasting. We covered it back up and did not attempt removal.

HOW MANY SUBS DO YOU NEED FOR THE SITE WORK?

The more contractors you have working on the property, of course, the faster the job will progress. This can save you money. On the other hand, if you prefer to have one person do most of the work you can wait for him. Normally, you hire one contractor to remove trees; one to dig the foundation, remove the tree stumps, and rough in the driveway; and possibly a third to install a septic system. The trenches for the water system may be done either by the earthwork contractor or through the plumbing contractor. Schedule carefully for all this; you do not want the septic tank sitting in the way of the grader or the tree cutters to spill over into the scheduling for the stump removers. This is why some people choose to just use one or two subs for the site work.

INSTALL TEMPORARY UTILITIES

Temporary utilities — water and electricity — must be in place before you can do anything else. The subcontractors will need them to do their work. Talk to each utility company well in advance for information.

PUBLIC WATER AND WELLS

If you are connecting to public water, a trench must be dug to the water supply and then the main water pipe is connected at the street. A temporary meter must be installed so that the water company can monitor the amount of water used during construction. If the main pipe bringing water in has pressure of greater than 80 PSI, it requires a pressure regulator. This reduces the water pressure so the pipes are not overly stressed — in the worst case, they may even burst without it!

If you are in a more rural area and public water is not readily available, you will install a well and septic tank. Both of these must be done by licensed professionals. First, the well company will drill down into the water table until they find clean ground water. Once they locate it, they install an electric pump, which brings the water to ground level for use. There will be a water storage tank installed so that the pump does not have to kick on every time you run a small bit of water inside the house.

Whichever type of water you install, be sure that the pipes do not leak at any location. Make sure all the connections are secured, and be sure that the spigot is flagged with a red ribbon so that no one runs over it.

SEWER AND SEPTIC SYSTEMS

Public sewer systems require cast iron or PVC pipes, which are laid in trenches before any concrete is poured. The sewer pipes must slope downward toward the sewer system. Be sure that you install an anti-return valve on the pipe, so that if the system overflows (this happens during heavy rainstorms) the water cannot back up into your house. Clearly mark the connection point, and after the pipes are properly buried have the inspector out to check the system.

If you are installing a septic tank, call a septic company out to choose the best location. The septic system is comprised of a tank and a leach field. The waste water from the tank drains into the leach field. The more bathrooms and bedrooms you plan to have, the bigger the required tank. Tanks of 1200 to 1500 gallons will service a three-bedroom, two-bath house. That may sound large, but the tank is buried underground and does not require a lot of room.

The tank does need to be positioned in a convenient spot so that a truck can approach it for cleaning. After the tank is in place, mark it clearly so it is not covered up during excavation. A septic company must inspect the tank once it is in place.

The problem, if there is one, will be with the leach field. It has to cover enough ground to absorb all the waste that the septic system sends into it. This is ultimately dictated by the type of soil that you have; a swampy soil may not be suitable for a leach field, a very wet soil will require a large leach field, and a fertile, friable soil will require a smaller field. You will have to have the soil perc tested to find out what size of leach field will be required.

There will be setbacks from the property line, and these include your septic tank and leach field. Be sure that your lot is big enough to accommodate all this and will still conform to setback requirements. Hire a waste engineer to plot it out if necessary.

Some localities are now requiring that you have enough room for two leach fields: the primary one that you are planning to use and a secondary one in case the first one becomes plugged. This is a real possibility, especially if you do not have your tank cleaned every few years. If your area requires the space for a backup leach field, make sure that your lot is big enough, otherwise you will not be granted the necessary permit.

ELECTRICITY

Unless you are going to live off the grid, you must run electricity into your home. This will either come in underground or from a pole. You can look at the property around you to see which way it is done in your neighborhood. The power company will run the line up to 50 or 60 feet at no charge, but if your site is very far from the pole it may cost as much as $25 per foot extra. You or your subs can do the trenching if it saves money and you can have a licensed electrical contractor install a temporary meter, but the utility company will do the connecting. In some areas, the contractor who installs the meter must be on the electric company's approved list. The building inspector will need to inspect your meter prior to the electric company connecting the power.

PUBLIC GAS AND PROPANE

Connecting to a public gas line is simple — the pipes are trenched to the gas line that runs under the street. For propane, you can get the supply company who will be delivering your propane to install the tank. But this is one item that can wait until later on; it is not necessary to have gas or propane until late in the building process.

Be sure that the gas pipes are properly buried, when it comes time. Clearly mark the connections.

EXCAVATION

Excavation should cover all areas of the foundation, fireplace, porch, and stoop. The crawl space should also be cut and graded in a way that makes sure it stays dry. The excavation process is one that you absolutely must watch. Do not be tempted to let the sub set this up without you. The whole time the work is being done, take depth measurements. If the foundation is too low, the sewer and water lines will have drainage problems; the sewer line might also be lower than its street side connection. Low foundations, as compared to the land surrounding them, are also susceptible to basement water problems. Water lines and footers must be placed well below the frost line, which can be two to five feet in the northern parts of the U.S. If the foundation is too high, you will end up having more front steps. Keep an eye on the work, and if it has to be a little off, let it be a touch higher than planned. You can expect all the excavation work to be within two inches of level.

While the excavator is working, have him clear the topsoil from the driveway, and pile all the topsoil where you can get to it later. Be sure that the excavator digs below the frost line, no matter where that is (you will need to confirm where the frost line is for your geographical area). Also, you can save money by having the water and sewer lines run in the same trench; go ahead and have this trench dug during foundation excavation.

Now install the batter boards, checking the corners for the correct angle as mentioned earlier. Batter boards are merely boards nailed horizontally to posts set at the corners of an excavation. They are used to indicate the desired level for the foundation. They also serve as a base for fastening tautly stretched strings that indicate the perimeter of the building. Here's how to lay them out: Drive three stakes outside each corner stake. Use sharp 2x4 stakes; make sure they are long enough to protrude six inches or so above the boards. Now, using boards of either 1x4 or 1x6, nail a board horizontally between a pair of stakes, say the top and bottom of the triangle. Repeat the process, placing another board from the bottom stake to the rightward stake. Lay the outside batter boards first, square them, then lay the inside batter boards. These batter boards serve to outline the foundation, leaving room for digging around them.

DRAINAGE

Now it is time to put a run of crusher stone onto the driveway. This will be used for a base for the paved driveway later, and it helps allow suppliers and subs to get in and out now.

The local government will not allow mud from your construction site to run onto the street or other people's property, so build a silt fence to keep the mud in. This is done with two-foot stakes and rolls of plastic or mesh. They can also be built with hay bales, which are then re-purposed for landscaping later.

FOUNDATION

The foundation phase is one of the most important parts of building your home. Hire the best contractors you can find. If you can locate a full-service foundation outfit, they will lay out the entire foundation, including the batter boards. They will then dig/pour the footings, pour the basement floor, and schedule all the inspections. Most of them even do the waterproofing. These companies are a little more expensive than hiring the work separately, but you will save in time and total costs, and you may even save some on your construction loan interest. They are worth checking out, especially if your plans call for a full basement.

If you are not going to hire contractors to construct the footers and foundation, take the frost line into account. If you happen to build the footers above the frost line, the weather will cause it to continually expand and contract. In turn, this can make the footer move, eventually cracking the footer and/or foundation. The purpose of a footer is to distribute the weight enough so that the house does not settle and nothing moves.

To pour a concrete foundation wall that will enclose a basement, measure a minimum of seven feet from the basement floor. Many contractors go up to nine feet. When back-filling the foundation, use the soil that was excavated to create the foundation — not the topsoil that was originally piled into a back corner of the lot when you first began construction.

CHAPTER 10 CHECKLIST

- Investigate local suppliers for building materials
- Prepare a materials list
- Get quotes from various suppliers
- Review the siting of the house on the lot considering
  - Neighborhood houses
  - View
  - Sun and weather
  - Slope of lot
- Review placement of batter boards for location, squareness and markings
- Place "Danger" signs and tape around the property
- Determine where materials and waste will be stored (preferably a significant distance from home site/work area)
- Clearly mark all trees showing which are to be removed and which will stay
- Remove trees and brush
- Grade the property
- Install temporary utilities
- Install the septic system

- Excavate as required
- Pour foundation — build up with block as required by the plan elevation

11

First Stages of Construction

FRAMING

Framing the house is always exciting because it suddenly looks like a great deal of progress has been made. After months of planning and working, it finally looks like a real house. You can see the plans you made on paper taking shape.

The framing of the house is absolutely critical. It is the support structure for the entire project, much like the skeleton of your body supports everything else. There is no room for error in framing; any deviation over 1/4" is considered unacceptable.

Framing Lumber

The lumber used for framing is kiln-dried, the higher grade the better. People have used green, freshly milled lumber for floor framing; as it dried it buckled, creating a wavy floor for the homeowners to walk on — not to mention an unsafe support system for the rest of the house, and a guarantee of shifting and settling later. Even if the lumber is seasoned but lower grade, it can sometimes shrink after it is in place. This will cause "nail popping" which is annoying and requires repairs. The better your lumber, the easier it will be for you or your sub to do a good job. If lumber is going to touch the foundation or concrete, it should be pressure treated pine, which will protect against moisture and termites.

If you have purchased a kit home, there will be precut materials that are possibly even preassembled. They will have their own set of instructions on how to proceed.

FLOORS

If your house is built on a concrete or block wall, bolts will be inserted in the concrete as it is poured for attaching a sill plate. The sill plate is a board which is then bolted to the foundation and becomes the anchor point for the floor and all exterior walls. Floors are framed to make a base for the floor to rest on. This support has to be strong and stiff, with no give that could result in settling. Joists are made of 2x8s or 2x10s, and are placed 12, 16 or 24" apart.

Floor joists that are excessively long are braced up by supports like steel I-beams. There are also vertical supports made of steel that rest on footings or piers. It is important to consider what the post is resting on; some builders place a wooden block and nail the post to it. Instead, embed steel plates into the concrete footers. Use these to bolt the post on. This provides strong support and a barrier against termites and other pests. Some contractors suggest using two boards fastened together, creating a double floor joist, around openings or spaces that are created for hot air ducts or bathtubs.

Sometimes instead of traditional joists, the boards are bridged (stabilized by using bracing placed diagonally between the joists). If this is the case, they are called trusses. These are becoming more popular, not only because of their strength but also because they make it easier to install HVAC and wiring, and because they tend to have better sound-reducing capabilities.

Subfloor

The subfloor is the platform where you attach the finished flooring and the walls. It is made from plywood or particle board and lies in the opposite direction of the joists. You may choose not to lay subfloors. If you can place the flooring straight onto the slab there will be no need for subflooring. However, the upper floors will still need subflooring, as will a house that is built on a beam foundation.

The subfloor is the main cause of squeaking floors. There are two steps that are necessary to prevent squeaks: the first is to glue the plywood down when building the subfloor. The second is that ring shank nails (the kind with screw threads) should be used to nail down the plywood. These will hold the wood down and will not draw back out, which is what creates room for the plywood to flex and squeak. Another alternative is to use joists made of composite instead of wood, so that they are not able to give.

You may have heard of a "floating" subfloor. This is thicker plywood flooring than the traditional 3/4". It is usually created from 1-1/2" plywood squares placed on piers. These are used-in areas where there is a lot of expansion and contraction, due to weather, soil, or temperature fluctuations. The floor can move slightly to accommodate the expanding and contracting. The finish flooring rests directly on top of the subfloor. If you have a floating subfloor, you will not notice a difference between it and a more solid floor.

There is one exception to the subfloor style, and that is a subfloor for tile in the bathroom and/or the kitchen. If you plan to use tile in these spaces, you must use a special compressed wood made especially for tile. It is laid over the plywood then screwed, glued, and sealed. Even better for tile is a cement surface. This is because cement can be leveled and smoothed perfectly. It can be strengthened with wire mesh and then the tile is glued directly to the cement.

WALLS

Once the subfloor is laid, it is time to build the walls. Wall framing has to be exactly right, so that there are no surprises, no changes, and so your house does not look like it was built by the "crooked little man" referred to in the song.

Making changes to the framing can cause many design issues, so consult with the architect one last time prior to taking on the wall framing. There are also some plumbing issues. One is that the framing should be spaced with room for the plumbing that will be put in. The other is that some shower stalls, especially the one-piece kind, may not fit through a regular-sized doorway once the framing is up. So go ahead and put the shower stall or Jacuzzi tub in place prior to framing. If you are concerned about theft, consider securing these in place with heavy chains.

Kinds of Wall Framing

Most framing in the United States is platform, or stick framing. This is where all the walls for the first floor are attached to the subfloor. The second floor rests on the ceiling joists for the first floor, and so on.

Another type of framing is balloon framing. In this type of framing, the vertical exterior walls extend from the sill plate on the first floor to the top plate of the top floor. These studs are more costly than the studs used for stick framing. The labor is also slightly more costly for the construction. Balloon framing requires that the entire outer perimeter of the structure be uniform. As a result, it is not always the best kind of framing for a given set of plans.

Pole framing can be used when strong winds are expected. Although it is not a common kind of framing, the poles are sunk deep into the ground and are exceptionally secure.

Post and beam framing uses rough-hewn wood to span long, open spaces. It does not use joists for the floors or the ceiling. This kind of framing is used for exposed beams and open ceilings.

Steel post construction is quickly becoming a standard in home construction; steel posts may be used at the corners of the structure, embedded in concrete. This is used in conjunction with walls that are filled in and do not support the ceiling above. For houses that are exposed to strong winds, metal hooks can be placed in the foundation with bolts that fasten to the steel rods; if the house moves up and down the wall will not separate from the foundation. Some builders run a metal brace on the top plate around the entire perimeter of the house as well, attaching it to the corner supports.

Money-Saving Tip:

The corners of the outer walls are required (by the local Building Code) to be braced. Using a diagonal wood brace cut into the studs, or a metal strap, which will save money and will allow for the insulating that is created by the sheathing.

Homes in the United States are built with 2x4 framing placed 16" apart. Stud grade lumber will cost more than standard framing lumber, but the studs will be uniform and straight, and will not have as many cracks. Whichever one you choose, do not be tempted to frame with utility 2x4s. These are uneven, making it nearly impossible to finish the walls properly, and they may not even support the load required.

Be sure that you have made plans for fire blocks to fit between the studs, midway between the floor and the ceiling. The function of a fire block is to keep fire from moving up through the walls to the next floor.

Before creating the framing, make sure that the placement of any openings will allow for good airflow. Better airflow takes place when air comes in and out through openings (windows) of about the same size. Make sure the windows are located in a way that allows for cross ventilation. Even though many homeowners today never open the windows, there will be times when you will want to air out the house, dry freshly cleaned carpet, occasionally enjoy fresh air, and so on. You will be glad to have good ventilation in these cases.

Interior Walls

The exterior walls support the weight of the roof. The interior walls, on the other hand, are created for partitioning the rooms and supporting the floors above. These walls are built in about the same way as the outer walls are — there are bottom plates, top plates, and studs spaced 16" apart. The biggest issue in interior wall construction is soundproofing them. Normal construction (2x4s with wallboard nailed over them) creates a hollow cavity where sound echoes and transmits easily to the next room. Since you are building or remodeling your own space, it is worthwhile to insulate the walls to cut down on the sound transmission. The cost for the entire home is only a few hundred dollars.

To insulate the walls, one partial solution is to stagger the 2x4 studs so that each side of the wall has studs that are for that side only; in other words, if you are working on an interior wall that faces the living room on one side and a bedroom on the other, every other stud will be positioned against the living room wall, with the others against the bedroom. After this is done, there is still a need for a further sound trap. This can easily be completed by filling the wall with insulation. Cut the insulation to fit exactly — do not fold it over inside the wall.

WINDOWS AND DOORS

It is important that windows and doors be framed properly, with double studs and headers that run across the top. The header's function is to support the weight above so that there is no sagging from overhead. Header sizes must be calculated carefully; it often takes strong beams to support all the weight.

SHEATHING

Exterior sheathing is made of plywood or particleboard. It creates a weatherproof base on which you will place the outer siding, bricks, shingles, or other type of wall materials. If you are going to use plywood, be sure it is exterior grade. It will be durable and will not weaken because of exposure to the elements.

Money-Saving Tip:

Plywood has become more expensive than ever, so consider using particleboard for houses that will have another exterior covering over the sheathing. If you are going to use particleboard, nail it twice as close as you would plywood.

ROOF FRAMING

The roof not only protects you and your belongings from the elements; if it is constructed properly, it will keep cool air and warm air where they belong, depending on the season. The roof needs to be sturdy — it has to hold up tile or slate, solar panels, skylights, and contractors walking on it, not to mention snow loads if you are in an area that gets snow! It also has to allow for items that can cause moisture to enter the home, like vents, skylights, valleys or dormers, and chimneys.

Most roofs are created from one of a few styles: gabled, gambrel, flat, hip, mansard, or shed. The steepness of the roof is called the pitch. It is a figure that is based on the amount of rise (vertical distance) in a given amount of run (horizontal distance). Pitch is described with the rise first: a 2-12 pitch means two units of rise for every 12 units of run. Normally the unit of measure is in inches.

PITCH	MEASUREMENT
Low	1-12 or 2-12
Medium	3-12 to 6-12
Steep	7-12 to 12-12

The pitch of your roof will depend on the amount of snow in your area, what kind of materials you are using for the construction, and how much space is beneath the roof. Climate and architectural appeal are also part of the decision process. A low pitch offers a roof that will shed water, but can hold snow — which actually adds extra insulation — in cold climates. Thinking of the space directly beneath the roof, there will be an area that feels open and has plenty of room for storage, with a sloped ceiling above. Interior maintenance may be easier with a low-pitched roof. This type of roof is less costly than those with steeper pitches because they require fewer materials. They can also be constructed so that you can hardly see them, which may eliminate the need for costly colored roofing tiles. Lower pitches also create a smaller overall cubic footage (length times width times height) meaning that the cost of heating and cooling is lower.

Roof framing will either be stick-built or pre-manufactured roof trusses. The trusses are extremely sturdy. They can hold up a large load and they do not require that there be load-bearing walls placed above the exterior walls, so there is more design flexibility. They can be installed in a single day. Roof trusses are also less expensive than stick-built roofing, because there is a savings in labor costs. The down side of roof trusses is that you will not have as much attic space to use for storage.

If you are creating a cathedral ceiling, you will probably use rafters instead of trusses. These are boards that cover the entire distance from the exterior walls to the peak of the roof, forming a skeleton framework on which the roof decking is attached. Rafter sizes are computed based on load-bearing qualities of the wood and boards used, the distance that must be spanned, and the pitch of the roof. In addition, the spacing, lumber grade, local wind forces, and amount of yearly snowfall will be a part of the computation. One of the best choices to use for rafter joists are engineered wood I-beam roof rafters. These are attached with special blocks. They are unusually strong and can span greater distances than most wood. The dimensions are precise, which means there will be no time wasted adjusting for size and no wasted material. They do not shrink or crack, and are lighter than most lumber. However, like all rafter roof framing, the load-bearing interior walls will be used for support.

ROOF SHEATHING

The roof sheathing is made of pressed chipboard or 1/2" CDX plywood. You should be sure to place plywood clips between each sheet to keep the roof's surface smooth.

ROOFING MATERIALS

The choice of roofing material affects many things, including the style and the look of your home. Let us look at some of the options.

Wooden Shingles (Shakes)

If you plan to use wooden shingles instead of typical asphalt or composite shingles, place furring strips on the roof spaced with 3" in between, cover the boards with roofing paper, then lay the shingles on top. The shingles will be able to breathe, so they are less apt to become covered in mildew. Wood shingles can be a fire hazard, so they are used less than many other types of shingles. They are also more expensive to purchase and install.

Ceramic or Clay Tiles

These are an especially attractive option for a Spanish or stucco house. Unfortunately, they are costly. They may be extremely heavy, up to 75 percent heavier than metal roofing. Consider the pitch of the roof and the load-bearing capabilities, especially if you are re-roofing an older home.

Asphalt Shingles

These are the traditional shingles used on homes. They are somewhat heavy, but they cost less than all other types of shingles. Most asphalt shingles carry a 15- to 25- year warranty.

Fiberglass and Composite Shingles

These are similar to asphalt in looks. They are flame-retardant and they have a long life. They come in a self-sealing version, which allows the heat to glue each shingle to the one below it over time.

Metal Roofing

Metal roofing is available in shingles and in sheets. It comes in many colors, is anodized (no more climbing on the tin roof to seal it), and it lasts longer than any other shingle. Most metal roofs are guaranteed for 30 to 50 years, compared to 15 to 20 years for a traditional asphalt roof. However, a metal roof can dent if someone walks on it or if a tree branch hits it. There are several types of metal to choose from. Aluminum is very lightweight, but must be coated to give it a proper finish. Steel roofs are normally coated in zinc or a combination of zinc and aluminum to prevent rust, then covered in an acrylic topcoat, which adds color and protection from the elements. They are the sturdiest option.

Copper roofing is expensive, but it weathers beautifully. Copper is soft and easy to work with — but that also means that it dents easily.

SAVING MONEY ON ROOFING

To save the most on roofing, be sure that the system you choose works well with the elements in your area. If your state tends to have lots of hail, for example, choose a roof that has passed Impact Resistance tests. Check with your insurance company; many will offer a premium discount if you select certain types of roofing materials.

Look at the heat resistance and cooling capacity of your roofing materials. A light-colored metal will reflect heat, which lowers your utility bills.

Choose a roof system that uses concealed fasteners that are made from a metal that is compatible with that of the roof. Some roofing systems use clips to fasten the panels together, rather than methods that put holes through the panels. Anywhere there is a hole, there is potential for water to enter the roof and ruin your home.

If you are remodeling and installing a new roof, consider using metal. It can usually be laid directly over the old roof system, eliminating a lot of time and waste disposal.

Roof Overhang

Although it may be tempting not to create much of an overhang in order to save on materials, there are several good reasons to create one. It will block the sun when it is high overhead, keeping the house cooler. In winter, it allows the sun in (because the sun is lower in winter) which allows the sun's rays to help heat the house.

The roof overhang also protects the sheathing, siding, doors, and windows from the weather. An overhang at an entry protects you from precipitation as you go in and out. An overhang above the window will allow you to enjoy the sound of rain falling, even with the windows open, without the concern that the rain may come inside.

The larger the size of an overhang, the less often moisture is found to be a problem, both on the exterior and on the walls of the foundation. Wet walls can cause mold and even lead to the rotting of wall studs. If your area has significant rainfall, use overhangs liberally. The optimum length of the overhang from the wall has to be calculated based on the angle of the sun in summer and winter (note that it is different at different times of the year), as well as the rainfall. Rain that is falling in winds of 30 miles per hour may fall at a 53 degree angle; if the wind is only 10 miles per hour, it falls at 22 degrees. So the typical rains and direction of the prevailing winds in your locale will have much to do with the way you create overhangs.

Flashing

Flashing is the material used on roofs around the chimney, vents, roof valleys, and any other joints or areas where different materials come together. Flashing protects the framing and insulation from the effects of weather. It should be applied in a way that causes the water to flow off the roof and not pool or pass through to the underlayment. If your flashing is inadequate, or if it is not applied properly, it can reduce the energy efficiency of your home and allow damage to the ceilings and walls.

When most people think of flashing, they are referring to aluminum. It is the most common material used, mainly because it is inexpensive and lightweight. Aluminum is also fairly resistant to corrosion. It is shiny, though, and it is hard to paint, so aluminum is not the most attractive protective coating to use if the roof is fairly visible. Copper is another option; it weathers to a lovely green patina, but it is more expensive.

If aluminum flashing is unattractive, perhaps vinyl would be a good second choice. Vinyl materials come in many colors and are not expensive. Since many soffits are also constructed from vinyl-coated sheets, it is possible to have a smooth look by incorporating vinyl — and you will also have almost no maintenance.

Chimneys

Flashing around the chimney is especially important, because the chimney is frequently a source of water damage. Water seepage can damage the roof sheathing and framing. This often occurs before you realize that there is a problem with the chimney flashing. The flashing that is placed between the roof shingles and the chimney sides is meant to keep the area watertight, preventing all water seepage.

To ensure the effectiveness of the chimney seal, it is best to use two layers of flashing. The first layer starts at the lower side of the chimney, working from the roof line upward. Some contractors take an extra step, wrapping these pieces into the shingles to create a woven solid layer. Over this, lay a second layer from top to bottom, arranging it so that it covers the top of the snug layer underneath. Seal the entire installation with high-temperature silicone caulk, especially the corners. Do not use spray foam-type caulk against the chimney. When the weather is bad or the roof shrinks, the double layer will keep the area watertight.

Occasionally a chimney is located in the line of fire, so to speak — right at the point where water is diverted from the roof. If your design involves this type of chimney design, consider creating a saddle from sheet metal that is shaped like an inverted "v." Install it against the chimney and seal it carefully, flashing at all joints. In addition to diverting water, it will keep debris and snow from piling up behind the chimney. This is especially helpful if you cannot see the chimney from the ground.

Gutters and Drainage

If water pools on the roof, it will work its way under the shingles. Water needs to be diverted quickly and in the most efficient way. That means in addition to the pitch of the roof, flashing will be added to channel the water down toward the edges. Gutters will then carry the water away from the house. If you allow the water to run directly from the roof to the ground, it will probably settle into the foundation of the house — giving you either a foundation problem or a basement flood.

Guttering can be made from aluminum, fiberglass, copper, or plastic. It channels water to drainpipes that are placed at each corner of the house. The guttering must slope down toward the downspout at a pitch of one inch for every foot, as well as slant slightly from the house so that even if the gutters overflow, the water will spill outward and not onto the walls. There should be enough downspouts to effectively drain the gutter in a hard storm. The drainpipes carry the water away from the house, either to the storm drain or just away from the house.

When choosing gutter, take into account the amount of roof that drains toward each one. Normal gutters are 5", but it is often necessary to bump up to a 6" to fully collect the volume of rain. This is especially true for broad, steep roofs.

There type of high-tech gutter keeps leaves and debris out while allowing water to run in. If you have a tall design that requires getting up on ladders to clean the gutters, you have many feet of gutter, or if you simply do not want to climb up there to clean them, you should consider installing these special gutters with a built-in gutter guards system. They cost more, but it is possible to have a savings in maintenance costs.

If these are not an option, consider installing gutters that have removable screens across the top, or removable caps on the ends. These are easy to take off if you need to hose out the gutter with a garden hose.

Downspouts come in several shapes: round, rectangular, and various corrugated styles. The shape doesn't matter but the corrugated kind handles freezing water more easily.

VENTILATING THE ROOF

Ventilation is one of the most important aspects of building a roof. Effective ventilation is necessary for many reasons: to remove odors and moisture, to remove hot air from the attic space, to cut cooling costs, and even to extend the life of the shingles. Without proper ventilation, the heat in the attic can literally cook the shingles. Good ventilation will also help prevent the formation of ice dams on the guttering and the roof channels.

If your plans call for an attic, you can simply install a vent at each end of the house. This will allow the moisture that accumulates in the roof to dry out. If you do not ventilate the attic, you will find that the temperature can become unbearably hot. Consider installing fans in the vents, which have automatic thermostats. These are extremely inexpensive, and they will prevent the house from overheating in the summer, plus it will extend the life of the roof. You should also have vents in the soffits to provide air flow from the overhang, up the roof to the end vents.

If there is no attic, use ridge vents to ventilate the roof. Normally a 1" gap is left in the plywood right at the ridge of the roof. The vent is installed in this gap.

EXTERIOR WALLS

By far the one part of your home design that is noticed the most is the exterior wall materials and finish. In fact, most people refer to their house by its exterior: "the painted beige house" or "the brick two-story." Thus, it is important to use a good scale as well as pleasing textures and colors on the outside of your home.

In addition to its cosmetic appearance, the exterior of the home offers one more protective layer from the elements, noises, and sudden temperature changes. Because of the expanse of the exterior, you have a chance to save thousands on your home's siding. Be sure to consider the cost of the materials, installation, and labor. Many sidings are available in numerous thicknesses, finishes, and colors. There are some that are installed horizontally and some that install vertically. There are many that, if combined together correctly, will not only look great but will also eliminate much of your outside maintenance.

Besides the look and durability, there are other points to consider when selecting siding:

* Weather resistance — whether it will fade or spot over time, its ability to bend during pressure from wind, hail, and snow, and moisture resistance

* Durability — resistance to scratching, chemicals, and its resistance to stray balls, tree limbs, and other objects that may strike it, and how much of a threat these are

* Fire resistance

* Sound absorption

* Insect resistance

Siding is a very personal choice, depending on the look you have in mind, and there are many to choose from. Here is only a brief overview of options and some of their pros and cons.

Aluminum Siding

The Pros

Aluminum is a low-maintenance wall covering. It has a baked-on finish and does not warp or crack. It can be manufactured in many finishes or textures. Some of the aluminum siding is designed to look like wood — painted wood, shakes, weathered barn wood, and much more. By selecting a thicker gauge of aluminum, you will get a stiffer siding that will sustain more wind and hard strikes by flying objects.

Aluminum panels are offered in both horizontal applications, which might have a width of 4 or 8" (like clapboard), or there may be vertical sheets that come in 10- to 16"-wide strips.

The Cons

Aluminum siding can dent if it is struck by an object. So if you have kids who play ball, or neighbors who do, you will want to back the aluminum with Styrofoam backing. Also, striking the surface can scratch off the finish so that the shiny aluminum beneath is exposed — hardly an attractive look. And, aluminum's price is based directly on the current cost of aluminum, which can vary wildly.

Vinyl Siding

The Pros

Vinyl is popular because it is so low-maintenance, strong and durable. It is also molded with color throughout, so scratching it does not cause a color change. To make it stronger and increase the insulating properties, it can also be backed with the polystyrene board like the aluminum siding. Vinyl siding does not dent and can withstand huge winds.

The Cons

Vinyl siding can easily buckle if the installation is done incorrectly. I saw a house with brand new siding — and ripples along every panel. The young, inexperienced crew the homeowner hired was not familiar with its use. Vinyl may also be difficult to install around unusual wood trim. It requires careful caulking at every seam, especially where it joins with other materials. Also, if aluminum (non-rusting) nails are not used, there will be rust smudges on the siding in a few years.

Wood Siding

If you are after a rustic appearance, consider using plank siding, like redwood, fir, pine, hemlock, cedar, spruce, or cypress. Solid wood siding is offered in many popular styles: beveled and beaded, v-groove, tongue-and-groove, and many other variations. Even the old-style clapboard siding is created from wood.

It is also possible to use plywood panels for siding. These are offered in 8- to 12' lengths and are very easy to install, which can be quite a savings. Plywood can also be applied over the wall studs without underlayment — a further cost reduction.

Hardboard sidings are those created from wood products and then manufactured as panels. They are stiff and look like authentic wood. These sidings have proven to be more dense than wood, and they do not split off into layers or crack. They can also be ordered already primed for the paint finish, making it easier to work with.

The Pros

Wood siding is inexpensive and gives character to a plain design.

The Cons

Every wood siding must be treated with weatherproofing. The coating will have to be repeated occasionally to prevent water damage, insect infestation, and erosion. This is true even of the pre-finished hardboard sidings. Wood also requires that a vapor barrier be installed underneath, to prevent condensation.

Brick and/or Stone

Brick or stone exteriors, or a combination of both, are beautiful finishes. They are durable and offer a positive selling point, as well as a good return on investment when you decide to sell the home.

The Pros

Brick and stone homes are sturdier when strong storms come; they do not sway in strong winds and will not crack. They are fire resistant and naturally repel a great deal of noise. They require the least of all wall coverings in terms of maintenance.

The Cons

Changing the exterior can prove to be a huge expense. It will be difficult to match the brick far down the road, and if you decide to remove a wall it can be quite costly.

Brick and stone are not strong insulators. Frequently, the homes covered in these materials are the hardest to heat. To offset this, you will have to use lots of insulation.

Because of the thicknesses of these coverings, the exterior walls should probably be moved inward 5" to accommodate their thickness. Also, the door and window openings will have to be reinforced with stone or metal lintels so that they can hold up the brick above.

Stucco

Stucco is a type of Portland cement that is applied with a trowel directly over the masonry or on a vapor barrier called stucco wrap. Stucco is applied in at least three coats totaling less than 1" in thickness. Although it has gotten a bad name in some areas of the U.S. because of poor application, if done correctly stucco is a sturdy and useful wall covering. Many home buyers feel that the "look" of stucco gives more value to the house.

The Pros

Stucco is easy to apply and lasts practically forever. It does not require maintenance unless mildew attacks; this can easily be removed with a sprayer filled with a combination of household cleaner and bleach. Stucco can be tinted with color so that it does not have to be painted, and the final coat can be done in one of several textured finishes, depending on whether you are creating a traditional or a more modern architectural style. If the stucco does crack, it is easy for a do-it-yourselfer to repair.

The Cons

On a wood-framed structure, it is imperative that a layer of stucco wrap (a type of construction fabric) is used. The wrap prevents moisture damage in the stucco by routing the water away from windows, doors, and joining areas. If this is not applied correctly, the stucco can crack. It will also crack if there are large walls without control joints placed at a distance of 3'.

Wood Shingles or Shakes

Wood shingles or shakes can be made from redwood, cedar, or cypress. They give a warm, rustic appeal to a home. Cedar darkens and weathers to a silvery color that many people find particularly attractive.

When buying shingles or shakes, you will see that the shingles have two smooth sides, while the shakes have at least one rough side. This is the difference in the two types. Higher grade shingles and shakes will not have knots in the wood or pockets of pitch, which can interfere with the application of lower grade shingles.

The Pros

If you are trying to create a rustic or woodsy look, you cannot go wrong with wood shingles. Wood is a good insulator and in some climates can be installed with no preservatives.

The Cons

Most wood shingles and shakes must be weatherproofed every five years. This can become a considerable expense over time. Wood shingles and shakes can be expensive, and often they are formed separately so that you have to nail them on one at a time. To prevent ruining their look, use strong, non-rusting nails.

To avoid some of the expense and time involved in covering your home with wood shingles, look for prefabricated 8' panels, or consider the vinyl or polypropylene versions of wood shingles. These panels stand up to the elements and are resistant to warping and twisting.

EXTERIOR FINISHING CHECKLIST

✘ Select materials for the house exterior considering

- Style of house

- Neighboring homes

- Material durability, weather resistance, insect resistance, etc.

- Cost

- Aesthetic qualities

✘ Install exterior materials

⚒ Caulk around all windows, doors, other openings

⚒ Paint or stain as required

PLUMBING

Plumbing System Design

For new construction, planning for the plumbing in your house is a critical task. While this may sound like a strong statement, it may be the most difficult single item to change once the house is complete. (A good argument could be made that it is second to the HVAC distribution system.) Therefore, the plumbing layout needs to be carefully thought out, drawn out, and then thought out again. If you are doing a remodeling project in a bath or kitchen, the plumbing modification can be relatively simple or very difficult depending on the extent of the project. The addition of a master bedroom suite with a bath will require tying in to the existing plumbing and this can present some difficulties. I will discuss remodeling further under the inside piping section.

When laying out the house floor plan, bathroom placement deserves some special consideration. Placing two bathrooms back to back so that they can share supply and sewer piping can save significant cost. The same thing is true by putting a second floor bath over a lower floor bath or kitchen. Also, one of the most commonly overlooked needs in plumbing design is outside faucets. Think about your needs for water outside and plan enough convenient faucets to serve all them. If you are going to have an unfinished basement, plan ahead for adding a full or half bath in the future and rough it in. It will cost very little now and will avoid a very large cost later if you decide you need one. It can even increase your resale value by giving future owners an option. As you think of the overall plan for making water available in your house at all the desired locations, there are three main components you need to consider. These are:

1. The water supply to the house

2. The interior piping to the desired locations

3. The waste system and connection to a sewer line or septic system.

We will look at these three subsystems separately.

Water Supply

This has already been discussed in Chapter 10, but it can be a major consideration which warrants another mention. The water supply to the house may come from a public utility or a private well. Most people would probably choose to use the public water supply if one is readily available, simply because it is the most convenient option and the one which leaves the responsibility of maintaining the operation and ensuring the purity of the water to others. However, in most areas drilling a private well may be an allowable option and, if you plan to live in the house 10 or more years, could save money over time. The decision of which to choose is one of personal preference and economics.

To make the financial determination you will need information on the average depth of the water table in your area, drilling costs, water quality, and treatment which will be required. Between the Health Department and the drilling companies, this information should be readily available. You will also need an estimate of your water usage to determine what your monthly public utility bill would be. Do not forget to include usage for lawn and garden watering. In addition, with a well you will typically need to build an enclosure to protect the pump and piping from freezing and for storing treatment and testing materials. Well systems will have to be inspected and approved periodically by the local Health Department.

Interior Piping

Interior piping for new construction and additions is done in either two or three stages:

1) Rough-in

2) Tub set and water pipes, often done at the same time as rough-in

3) Trim out

Rough-in includes locating and installing the flanges and sewer lines for all toilet fixtures as well as the main sewer lines and branch lines within the house. If there is going to be concrete poured for the first level or basement and sewer lines are to be encased in the concrete, the pipes must be put in place before pouring the concrete. All sewer and drain lines must be vented properly to allow smooth water flow. If this is not done you will get the same effect as turning a gallon jug of water upside down — a lot of gurgling and chugging, but not much water flow. The venting is normally done through a pipe tree (or standpipe) which accepts drain lines from several sources, connects to the sewer line below and vents up through the roof. These are typically constructed by the plumber depending on the configuration of the sources.

There are a couple of hard and fast rules. Every drain line or sewer connection must be vented to the outside via a standpipe with no obstructions. Each line connection to the standpipe must have a water seal between the vent pipe and the source. (This is the S-shaped pipe you see under your kitchen and bathroom sinks.) Do not allow a sewer standpipe to vent into the attic because you do not like the way vent pipes break the roof line of your house. You will get, at best, excessive humidity with accompanying mold and mildew in the attic, and probably a house that has a distinctive air about it. Sewer and septic systems generate noxious and flammable gases during the decomposition process. These gases need to escape and the vent lines are that escape route. If you fail to install a water seal (trap) or if you terminate the vent in the attic, your house will become the recipient of these unsafe and unpleasant gases.

If there is a second story with a bath, the sewer lines will have to be routed down through a wall. The vent lines throughout the house must go up through the roof, through a wall. Therefore, these lines cannot be installed until the walls are framed in. The water lines also have to go through these same walls. For this reason, it is not unusual for the rough-in and water lines to be done at the same time, immediately after the framing is done. The tub unit, especially a one piece shower, will often be set into place even before the interior framing is complete just for ease of movement through the house. Since interior sewer lines from upstairs sources have to come down through the walls, some extra consideration must be given to noise: Who has not heard that middle-of-the-night gurgle from someone upstairs flushing a toilet as the water runs down through the pipe? Fifty years ago all the interior sewer lines were made of cast iron. Since then, some much more economical and easy-to-install materials are being used (more on materials later). Cast iron is very heavy and does not transmit the sound nearly like the modern materials. You should avoid dropping the sewer through a bedroom wall where possible. If this is not possible, then some insulation packed around the line inside the wall will offer some sound reduction. All water lines must be run and stubbed out before sheetrock or paneling is done.

Since all the sewer or waste water flows by gravity, the interior sewer and drain lines must be sloped toward the point where they leave the house. The rule of thumb is 1/4" per foot. If this is not followed there will be trouble. No cheating allowed here. Sewer pipe hung between the floor joists must be properly supported and sloped. Be sure to check the piping manufacturers' recommended minimum distance between supports for the specific diameter and schedule pipe you are using. In addition, while water lines can be put through holes drilled in the floor joists, sewer lines cannot pass through a floor joist. The maximum size hole drilled in a joist is one-third the size of the joist and all holes must be a minimum of 2" from the top and bottom edges of the joist.

Trim out is the final stage. It includes setting of toilets, installation of faucets after sinks, lavatories, and putting countertops in place. Tiling work has to be completed in showers and baths. The supply water and sewer connections should be complete before the trim out is done so that the system can be tested as areas are completed.

Pressure Reduction

This is a good time to talk a bit more about the pressure reduction valve (PRV). It is a very good idea to locate it an easily accessible location (part of the plumbing design). They do need readjustment occasionally and will sometimes get clogged with small particulate matter from the water lines. Many people like to locate the take off line for the outdoor faucets ahead of the pressure reducing valve to take advantage of the higher pressure for outdoor use. This is fine so long as the pressure does not exceed about 110 psi. Much more than this and you may find your garden hose life significantly reduced. In addition, if the water system pressure exceeds 120 psi this could cause another problem. The water piping most often used to connect your house to the water system is available with ratings of 100 psi and 125 psi. Depending on which you select, you may have no choice other than putting a PRV immediately after the water meter. You would still have the option of having increased pressure for outside faucets by adding a second PRV inside the house after the take off for the outside faucets. You could then set the PRV at the meter at 100-110 psi and the inside PRV at 80 psi. The PRVs are relatively inexpensive and I find the increased pressure outside to be well worth the cost. Most hardware stores, particularly the superstores, sell a relatively inexpensive water pressure gauge which screws onto an outside faucet or clothes washer faucet to allow you to set the PRV correctly.

Sewer System

The third component is the sewer system. A fairly complete discussion of the requirements of system was given in Chapter 10. If you are not able to connect to a public sewer system and must install a septic system, it is worth repeating that it is best to obtain an approval on the size and location requirements early in the process — before you finalize the layout and grading plans for the house. Sometimes the percolation tests can turn up unexpected rock and force the placement to a less than optimum location. The time to discover this is during pre-construction, not after the foundation is poured.

All sewer systems depend on gravity to work. Waste water must flow from the drains and toilets in the house, to the sewer line or to the septic tank, and then to the leach field. There are situations where gravity needs a little help. Some houses are located at an elevation below the sewer line or below the required location of the septic system. In these cases you will have to install a sump below the house elevation with a pumping system to pump the waste up to the sewer line or septic system. These pumping systems are readily available at Home Depot, Lowe's, and other places, and are reasonably priced. Do not despair, this is not a show stopper.

As you can see, the water supply and sewage systems for your house have major impacts on house placement on the lot, landscaping, and cost. Proper planning and timely approval are essential to avoiding costly mistakes.

Natural Gas and Propane Piping

This has been singled out for comment because fuel piping is an area which deserves special attention. Piping for natural gas and propane, as well as the installation, is very strictly governed by Building Codes. The materials used (normally black iron pipe and copper tubing) are specified as to grade and thickness and precisely where and how each is used. This is work which must be done by the gas company and/or a licensed installer, usually a combination of the two. It will be done under the close scrutiny of an inspector who will witness a leak test on the system and many other details. A mistake in these systems can cause the total loss of your house and can kill. Leave it to the experts. It is a minor price to pay.

SAVING MONEY ON PLUMBING

There are a number of opportunities to save money on your plumbing. Some involve material selection, which can save on building costs in terms of material and labor. Before we get into materials, we should look at some decisions which can save on an ongoing basis. The most expensive monthly cost related to a plumbing system is hot water (you might rightly consider this an energy cost as well). The obvious place to look first is the hot water heater. Whether you have a gas or electric heater, its sole purpose is to heat water to a predetermined temperature, hold it there until needed, then repeat the process. Water heater efficiency has been improved substantially over the last five years. All have energy guide labels which will help you pick the most efficient. The labels also contain information on the family size rating for the different heater sizes. Do not buy a larger heater than you need. Look also for heaters that allow you to program a heating cycle, allowing a lower temperature during the day but having the water hot for when you get home.

Consider heater placement in your design. Try to locate it as close as possible to major usage areas like the laundry, kitchen, and bathrooms — perhaps at a central location rather than a garage on the opposite end of the house. Be sure your plumbing plan includes a floor drain near the heater, or if it is in an interior location put it in a pan (available from the same sheet metal shop that does your ductwork) with a drain in it. This is an inexpensive insurance policy should the pressure/temperature relief valve or any of the connections leak unnoticed. One of the biggest uses of hot water, believe it or not, is going into the kitchen, turning it on, and letting it run until it gets hot when you need only a small amount. One great option is an instantaneous hot water heater which mounts under the kitchen sink and gives you up to 60 cups per hour of hot water.

When selecting your faucets and shower heads, new low-flow designs are readily available. These will reduce the amount of water used by these devices while still allowing them to perform their desired function adequately. Low- flow toilets are the only option available now, so selection of these can be based on more aesthetic qualities.

MATERIALS

Normally the first thing most people think of when they hear the word plumbing is pipe or tubing. This discussion has been saved for last because of the sheer volume of choices. When determining what materials you want to use for piping the determining factors are: material cost, installation cost, any health considerations (approved for drinking water), life expectancy and pressure rating.

If you are a do-it-yourselfer, you may be confused by the varieties of pipe available on the market today. Personal preference is always a factor, but one must be open to many of the improved materials available today. The choices come down to a reasonably straightforward decision for any given use. Each pipe has its own particular use, which is often dictated by plumbing codes in your locale. Climate can also be a factor; living in a temperate zone means your needs are different than a zone that has days of below zero weather. There is also the concern about health hazards in the home. This section focuses on the types of pipe and tubing materials available, and their applications. The first thing to recognize is that there is no universally correct answer; different applications call for different types of pipe.

MATERIALS AVAILABLE

Here is a list of plumbing pipe and tubing materials that are most commonly used today:

- Rigid Copper
- Flexible Copper
- Polyethylene
- Galvanized Steel
- Rigid Plastic PVC (polyvinyl chloride)
- CPVC (chlorinated polyvinyl chloride)
- ABS (acrylonitrile butadiene styrene)
- PEX (Crosslink Polyethylene)
- Braided Connectors

Let us look at these a little closer. Each type has its own special attributes and usage.

Copper has been the number one choice of plumbers for decades and would probably still be considered their choice for main water lines to the house, hot and cold water supply lines in the house, and running water supply lines and connections under sinks in the bathrooms and kitchen. Coming in two types, rigid and flexible (rolls), it is a versatile material. The connection and joining method depends on the type of copper. There are two basic designations — rigid and soft. The following considerations should be a part of the decision making process before deciding on copper.

1. The price for copper has been rapidly rising. It has gotten so high that some people have recently been arrested for stealing high voltage wiring (in service) to sell for scrap. The high cost has quickly been reflected in the price of copper pipe and tubing.

2. One of the concerns you have to deal with in house design (depending on your location) is freezing of the water pipes. While the choice of material will not determine whether the freezing occurs, different materials react differently to the freezing process. When water freezes it expands. Even rigid copper is somewhat malleable (which means that it can be deformed to some extent). Unless it is severe, it can usually expand and survive the first freezing without bursting. However, copper work hardens as do most other metals. If you are not familiar with this term, the easiest way to demonstrate it is to take a paper clip and bend it back and forth. The first several times it bends easily but with repetition it gets more difficult until the metal hardens, becomes brittle, and breaks. Some of the other types of materials will withstand repeated freezing better than copper.

3. When putting in the main supply line to the house, the pipe is necessarily buried in the ground below the frost line. If the soil is rocky or has a number of small rocks or pebbles, these will rub against the surface of the copper pipe as the pipe expands and contracts, eventually wearing holes and leading to leaks. If your soil is likely to have this issue, seriously consider an alternative to using copper.

Rigid copper has been widely used since the 1960s in home construction, and for many years before that in industrial applications. It comes in straight lengths (joints) from 5' to 12', in various diameters and wall thickness, and is suitable for pressure. Rigid copper pipe cannot be bent to go around corners so all fittings (ells, tees, caps and threaded connectors) are normally soldered to pre-cut lengths of pipe. Soldering, or sweating, is a technique of joining two pieces of metal using a low melting point (180-190 degrees F) fusible metal alloy. For years the most popular alloy for copper soldering was a tin-lead mixture. Since 1988 lead has been banned in drinking water applications and normally a tin-antimony or silver alloy is used.

Soldering is a technique which can be quickly mastered by the do-it-yourselfer. The secret is that the two pieces being soldered must be clean and dry, and acid flux must be used in the connection. The connection is then heated with a torch until it is hot enough to melt the solder. At this point, the flux sweats out of the joint and touching a piece of solder to the joint will both melt the solder and pull it into the joint. Crimp connectors and compression fittings are also sometimes used in plumbing applications, but soldering is the most common method for houses. It resists corrosion and lends itself to many different applications.

Soft copper is supplied in rolls. The two advantages it has over the rigid copper are that it is easily bendable (to get around corners and obstacles and up through walls) and that it comes in longer lengths which require fewer fittings. It can be flared for connector fittings, unlike rigid copper pipe, but can be soldered. It is commonly used in attaching water fixtures, faucets in the kitchen and bathroom, and also for hooking up dishwashers and ice makers. It is sometimes used in refrigerant lines and HVAC applications.

Note on refrigerant and HVAC usage: The EPA now requires anyone installing and servicing HVAC equipment to pass a certification exam. The test and training to take it are fairly expensive, and the failure rate is fairly high — so the do-it-yourselfer may be out of the picture here.

Rigid PVC pipe is available, conforming to standards for both pressure and non-pressure applications. It cannot be used for hot water. It is used for interior water lines, water mains, sewers, drain lines, and vent lines. It is very durable and is resistant to many ordinary chemicals such as acids, bases, and salts. It comes in 10' lengths and, like rigid copper, cannot be bent to go around corners. There are a wide variety of fittings (such as ells, tees, caps, reducers and threaded pipe connections) available. The pieces are joined with solvent cement which is easy to use. It is much less expensive than copper and is being used for many water supply installations.

ABS pipe is easier and less expensive to install than metal piping. It features superior flow due to smooth interior finish, does not rot, rust, corrode, or collect waste. It can also withstand earth loads. It resists mechanical damage, even at low temperatures, and performs at an operational temperature range of -40°F to 180°F. ABS has many of the same physical attributes as PVC. It is used primarily in venting, waste removal, and drains. While it is available with a pressure for industrial use, it is used only in non-pressure applications for houses. It is generally black.

PEX piping is not without controversy and is relatively new in the United States. It has been used extensively in Europe for over 35 years, but in the United States many wonder about its uses in certain applications. PEX is made from high density polyethylene. Its uses are in hydronic radiant heating systems (where hot water is circulated through pipes to a baseboard, radiator, or in-floor tubes), domestic water piping, and insulation for high-tension electrical cables. One of its major advantages is its flexibility, making it perfect for difficult applications. PEX is often installed using a distribution manifold which connects to the main water supply. This makes it very easy to work on specific areas of the home. For example, you could turn off the water at the distribution manifold to the kitchen sink and make repairs without shutting off the water supply to the rest of the home. If used in a hydronic radiant heating system along with ferrous materials, it must contain an oxygen barrier to avoid rusting of the ferrous materials. This is accomplished by sandwiching aluminum tubing with the PEX. It typically has five layers. PEX uses specially designed fittings and requires a special crimping tool. PEX is now available in the hardware superstores, and many hardware stores and rental shops are now renting the crimpers. The materials are priced below that of copper and the installation time is reported to be half that of copper's.

Galvanized steel pipe is generally not used in plumbing houses any longer. It is typically found in older homes and therefore may crop up in a remodeling job. It is no longer used due to its tendency to clog with mineral deposits and scale, causing high pressure drops in the line resulting in low water pressure. It is also subject to corrosion after the surface has been scratched, exposing the steel to the elements. It was regularly used for buried water supply lines to the house. It is still sold in the standard 21' lengths. It is cut to length using a heavy duty tubing cutter or hacksaw, and then threaded for the application. Joints are all screwed and use pipe thread compound to insure a tight water seal. Standard fittings are available, such as tees and elbows. Galvanized pipe is now often used for railings or replacement of existing galvanized pipe in the home. Installation is labor intensive and due to the difficulty in getting proper fits with making each screwed joint, best left to an experienced pipe fitter.

CPVC pipe is suitable for hot and cold water distribution. It has a 400 psi pressure rating at room temperature and a 100 psi pressure rating at 180 F. It is resistant to many common household chemicals, is corrosion resistant, and is widely accepted by codes. CPVC pipe and fittings are assembled with a solvent, the same way as PVC, but the solvent for one will not work on the other. CPVC pipe used for drinking water may leave a taste in the water for the first year.

Polyethylene comes in small tubing sizes and in pipe sizes. It is available in rolls of 100' and is pressure rated at 160 psi. It is approved for drinking water use and is therefore another candidate for your main supply line at 20 to 30 percent the cost of copper. The tubing sizes are good for such things as ice maker connections.

Braided connectors, while often shunned by the professional plumber, are the answer to the do-it-yourselfers' prayers. These connectors are braided PVC tubing reinforced with an internal woven fabric or external stainless steel mesh. They come with threaded connectors already attached on the ends in sizes suitable for their specific use. They are used in connecting faucets and toilets to the shut-off valve on the water supply. They come in a number of lengths and are close to foolproof (nothing to do with plumbing is absolutely foolproof). For the do-it-yourselfer who does not want to invest in flaring tools (much less learning how to use them) or swage fittings, they are a great option.

Choosing pipe to use for any application is important. The use of specific materials is dependent on local Building Codes. It is advisable to check with local building officials and inspectors before proceeding in any application. Using the proper pipe for any typical application makes the job easier, and checking with local authorities makes the application up to code.

FIXTURES

It has been mentioned several times that one of the first things to do is establish a budget for your house or remodeling project. There are a few places where you will be tempted to violate that budget and one of those is in the selection of plumbing fixtures. There are a wealth of styles, colors, features, and prices to tempt even the most determined building materials shopper. There are a couple of places you might want to push a bit for better quality and a couple where you can settle for less. We will look at the major areas.

Tubs and Showers — How many modern houses have you been in which have large Jacuzzis in the master bath suite that are covered with dust? These always sound like such a good idea, a giant bubbling pool of soothing hot water with water jets massaging your tired muscles. However, when you have figured out the water and electric bill increases, it does not sound quite so inviting. Besides, who has the time to relax in one? Before you pick one of these, ask around among people who have one and find out how much they use it. If you still must have one, consider one the size of a regular tub. It will cost less to buy, install, and operate. Assuming you have gotten past this and are just trying to pick out a bathtub, decide what size and features you want. Do you want cast iron or fiberglass? While this is somewhat a personal decision, the enameled tub should be more durable long term. If you go with fiberglass; be sure to check the stability of the bottom of the tub. Step inside one. If you feel the bottom give when you step around, step out and walk away. If it is flexing now it will continue to do so until you have a crack and possibly a leak in the bottom. While fiberglass can be repaired, a flexing repaired tub or shower will break again. If you are remodeling and putting in a fiberglass shower, you will need to look for the remodel styles and sizes. These are made to get through the house and into the existing bath. Remember one thing: Once this tub or shower is in, it is probably there for the duration. Do not buy less than you want with the thought of going back later and replacing it. It will cost too much to do that. Buy what you have to have now and save the money elsewhere.

Vanities and Sinks — while the vanity is not technically a part of the plumbing, it holds part of the plumbing (the sink) and the plumbing helps to hold it in place. There are almost as many choices as there are people to make them. Tops come as pre-rolled plastic laminate, marble, granite, tile, and a host of others. Sinks can be one-piece construction with the countertop or a drop-in. When considering the material you have to weigh initial cost versus durability. This is especially true in the kitchen where the countertop gets a lot of wear. Pay particular attention to the edge of the countertop and realize how many times a day something will be dragged across it. While not as difficult to replace as a tub, a countertop, especially one with a molded in sink, is a major expense. Try to get the best quality you can while meeting your design requirement from an aesthetic standpoint and staying in your budget.

Toilet — All toilets are now required to be low-flow designs so you are primarily going to be choosing based on style and color. The cost range variation is not excessive across the basic product lines, but there are some very high end products are available. They all adequately perform their required function.

Faucets — This is one of the places where you can let your artistic side get the better of you. Faucet sets for the bath can run from under $100 to over $1000. They all do the same thing … start and stop the flow of water into the sink. Granted, some do it with a lot more class and style. In the bath you have an extra choice to make: standard or wide set. Typically, the wide set ones are more decorative and more expensive. This is an area where you might think about going cheap now and replacing later. This is not too outlandish an idea as swapping out a faucet set is about a two hour job for a do-it-yourselfer. A lot of the cost differences between the top and bottom line are in the design shape and the finish applied. The basic function of all is relatively the same.

Water Heating — This has already been mentioned to some extent under the plumbing savings heading. The average superstore will have a good selection of water heaters in various styles, sizes, features, and efficiencies. Your first step is to select the size based on your family size and how you use the water. If you are building a large house for a small family, you might want to base the size on the number of people who might live in the house, just for resale. Then look at the warranty, energy usage, and features. Manufacturers are beginning to put a lot of controls on heaters and the features determine the price range. Do not pay for more than you want or need.

PLUMBING CHECKLIST

- Review plumbing layout plan to make sure hot and cold water are available in all locations, inside and outside
- Determine what materials you will use for all plumbing applications (hot, cold, pressure, non-pressure, supply, and sewer)
- Decide whether you will connect to a public water supply or well, then connect and provide temporary water to the site

- Install rough-in piping, including placing large one-piece showers and large tubs or Jacuzzis
- Make sure all vent pipe trees vent through the roof
- Make sure all sewer waste lines slope toward the outside of the house and are properly supported
- Install pressure relief valve on the supply line if required
- Choose fixtures for the bathrooms and the water heater
- Have HVAC piping installed through framing as required
- Have rough-in inspected before walls are finished and cover the work
- Do final trim out after walls are finished and cabinets are installed
- Have final plumbing inspection (if required)
- Electrical

CAUTION

While it may look like wiring a house is a simple thing, it is not. This work should, and in many cases must, be performed by an experienced, licensed electrician. There is more to it than meets the eye. Configuring circuits, proper wiring to common and hot sides of the panel, grounding, and GFCI installation are critical. One wrong wire can lead to fire or death.

PLANNING

Planning for the electrical system, switch and receptacle locations, lighting, appliances, and so on are very critical. Proper planning up front can make living in your house more enjoyable for years to come. Before calling the electrician to come to the construction site and start the wiring, you need to have considered the layout and needs for your home. I will discuss this more as I discuss each element.

Your electrical contractor is responsible for all wiring beyond the meter box. The power company is responsible for providing power to the meter box. In some areas, the power company is responsible for installation of the meter, but in others it may be done by an approved, licensed electrician. This will not be done until all electrical inspection has been completed and approved. It is important that no electrical wiring be covered or enclosed prior to the inspection. Electrical inspection is normally done twice, once after rough-in and again before the power is connected. For any new construction, local building codes will be very strict concerning wire types, wire sizes, breaker types and sizes, junction box sizes, receptacle box sizes, Ground Fault Circuit Interrupter (GFCI) locations, and many other components. Your contractor should have a good working knowledge of the local codes and know all of the requirements. You need to work closely with your electrician to make sure all your needs are met and within the code requirements.

The rough-in electrical work starts after the rough-in plumbing and interior piping have been completed, and the HVAC units and ductwork are in place. This order ensures that wiring will not have to be moved to allow for plumbing and ducting, which are not flexible. Rough-in includes locating the main breaker panel and installing a box for every receptacle, light, appliance, and switch, and then running the wire between them. No connections are made at this time. The wires are all left hanging out of the boxes. The finishing electrical work will involve wiring in all switches, receptacles, lights, connection of the dishwasher, central vacuum, any built-in appliances, fans and hoods, HVAC, etc.

MAIN SERVICE PANELS (LOAD CENTERS)

In the initial planning for the house you will have made several decisions which will impact the electrical system. Are you going to have overhead or underground power supply? Will you have gas, oil, solar, coal, wood, heat pump, or electric heat with air conditioning? Are you planning a heated swimming pool, hot tub, or a workshop with power tools?

All of these will impact the amount of power which will need to be supplied to your house and therefore the size of the service, which includes the wiring size coming from the power company and the size of your power panel(s). The main service panel (sometimes referred to as the breaker box or load center) is the distribution center for electric power to your house. It should be located very near the point at which the power comes to the house. The wires from the meters come through a conduit and attach to three large metal bars called busses. Circuit breakers clip onto the bus and the wiring is connected to the bus and the breakers. Building Codes will govern the minimum service you must have (rated in amps) but you may need more depending on your answers to the questions above. The most common size panel for new construction is 200 amps. You may need more than one. They come in various configurations and will accommodate breakers of various ratings. Circuits feed out of these panels for both 120v and 240v applications. The electrician will normally determine which loads go on which circuits so as to not exceed the breaker rating. However, if you have special requests they should be able to work with you as long as it makes sense electrically. Grounding for your electrical system is made at the panel. A heavy copper wire will be run from the ground bus in the panel to a copper rod driven into the ground (hence the name). Years ago the water lines were used as a primary or secondary grounding system. This worked fine when the pipes were all copper or iron and every joint had metal-to-metal contact. It is not an option with today's non-conducting products used for water lines.

A part of the electrician's job is labeling all the breakers in the panel so you know which breaker controls which circuit. These lists are sometimes not complete and taking the time to make sure you have a complete list can save you a lot of aggravation in the future (and probably in the dark).

CONDUCTORS

The minimum size (gauge) of the wire in your house will be determined by the Building Code. You may, however, want to use a larger wire size in some instances (remember that the numbers seem backward, 10 gauge has a larger diameter conductor than 18 gauge). Why would you want to do more than is required by Code? Have you ever noticed that when the hair dryer is started the lights dim? This is because the requisite surge in power exceeds that which the wire can carry at the available voltage and the voltage drops slightly until everything is in balance. It is like having your own personal brown-out. It is the electrical equivalent to someone flushing the toilet when you are in the shower.

The longer the run of wire from the panel, the more the total resistance and the higher the chance of voltage drop. The fix for this is to use larger gauge wire for the long runs. As with all electrical decisions, now is the time to do it. The difference in the cost of wire size now is small compared to trying to change it out later. Virtually all wiring used today has three conductors which allow all fixtures, switches, and so on to be grounded through the ground rod and also makes possible the use of GFCIs.

SWITCHES

"Where's the light switch?" If you have ever asked this question someone did not do their job. The Building Code will again regulate some parameters such as distance from the floor, but you can have input in regard to how many switches you have and their location. You should review the electrical layout drawing carefully. Think about entering and exiting the room and make sure there is a light switch next to the door, just inside the room. Remember that the boxes for switches will be installed when the walls are only framed. It will not necessarily be obvious at that point on which side the doors will be hinged. A switch behind the door is not the optimum location, but a lot of them end up there. Consider where you will want three-way switches (hallways, rooms with two doors) and dimmer switches. You may want some wall receptacles to be switched so that the switch will turn lamps off and on. One thing often overlooked is switches for exterior lights. You might want a switch in the bedroom to allow you to turn on the outside floodlights without walking to the other end of the house in the dark. Would you like to be able to control the bedroom lights from the bed? Now is the time to plan for that. Time spent studying and determining switch placement will be well spent.

There are several different types of switches you can choose from. The standard is the toggle switch, which normally comes in white, brown, or ivory with matching cover plates. You can get switches that click and silent switches that do not. Also available are rocker switches and a number of different decorative styles. Switches can also be lighted, which is not a bad idea for a bath at night. Dimmer switches are also available in a variety of colors and styles, including several remote control styles. Let your budget be your guide.

RECEPTACLES

While it may be possible to have too many receptacles, it is highly unlikely. Code requires a minimum of one receptacle on each wall and placing them at specified intervals. You might want more. In the kitchen, think about what small appliances you use daily and in cooking. It is also a good idea to have receptacles on several different circuits in the kitchen. For instance, put one on the circuit with the microwave, one on the refrigerator circuit, and one with the dishwasher. This will allow you to have several appliances going at the same time without worrying about tripping a breaker.

Plan for where your entertainment center will be, along with all its associated equipment, and plan for extra receptacles there. If you are going to have a shop or workspace in the garage, you will need to have enough receptacles to run all your tools. You might want to give consideration to having several separate circuits there as well. Do not forget your needs outside the house for everything from hedge trimmers to Christmas lights. Weatherproof receptacles and switches are required for outdoor use.

All outdoor, kitchen, bathroom, and garage receptacles are required to be on Ground Fault Circuit Interrupters (GFCI). These are devices that will interrupt the flow of electricity (trip) when a current-to-ground is detected. Their intended function is to prevent electrocution. They are required by Code. There are two types which can be used. You can get a circuit breaker which has the GFCI built in. Once installed in the panel it will protect every receptacle fed from that breaker. The other type is the GFCI receptacle. The receptacle can be wired to provide protection for other receptacles as well. All GFCI devices have test buttons which, when pushed, create a ground fault and trip the breaker. They should be tested frequently. If they fail to trip, they should be replaced immediately. They should be replaced by an experienced electrician because if they are not wired correctly they will not work as intended, leaving you unprotected. All GFCI receptacles are required by Code to be marked.

MISCELLANEOUS WIRING

There are several other wiring options which should be considered at this time. If you are going to have a security system, the security company should do their preliminary wiring with the other electrical rough-in. This is typically included in the estimate from the security company. Other wiring which the electrician will do for you includes cable and internet outlets, telephone, any speaker systems or intercoms which you want throughout the house, and home theater or surround sound wiring. When laying out these systems, think ahead to possible future needs. An additional cable connection will cost very little if done now. In addition, do not forget doorbells, the HVAC thermostat, and central vacuum outlets.

LIGHTING

You will recall the mention of several areas which can be budget breakers. Lighting can certainly be counted among them. While it is certainly possible to buy a ceiling fixture for under $10, it may be impossible to find one you like, especially among the host of more decorative and aesthetically pleasing ones. Ceiling light (non-recessed) replacement is another easy do-it-yourself project so you might want to go low cost for now with an eye for replacement in a few years, allowing you to save your budget for the major light fixtures like the dining room and foyer chandeliers. While the more ordinary lighting fixtures can be had quite reasonably, there are a lot of them required for a whole house and it can add up quickly. Since you have created your own budget, you will have hopefully spent some time looking at lighting fixtures ahead of time and budgeted accordingly.

HEATING, VENTILATION, AND AIR CONDITIONING

In your initial planning for the house, you should have gathered data and cost information to help you determine what options you wanted for your HVAC system. Much of your decision will have been based on the climate in which you live and fuel and energy cost comparisons for your area. The first fundamental decision is whether you want heat and cooling. In some areas you may not need cooling. In some you may have only a minimal need for heating. In most cases you will have a need for at least some degree of both. You can choose to have entirely separate systems, one that shares some components such as the fan and ductwork, or a completely integrated system like a heat pump.

Much of what you do will depend on energy availability and cost in your area. Which is more economical: natural gas, propane, fuel oil, electricity, or some other fuel? You may want to use wood stoves or fireplaces for heat or even solar heating. To make this decision, you will probably want to talk with a heating contractor and/or your local gas, oil, and electric companies. Many times they will offer consultation without charge to help you determine the capacity of the system you need, its installation cost, and an estimate of the operating cost. Often, the gas and electric companies will offer a discount on the systems and appliances or low cost financing if you will commit to their energy source for your HVAC and water heating. Check the energy ratings on the various systems. For two-story houses or houses over 2500 sq. ft., you may want to consider a zoned system that uses two smaller systems instead of one large one. One good indicator is to see what is being used in most new and remodeling construction in your area. If you are used to having gas heat and are considering a heat pump, remember that the heat pump circulates heated air that is only 90-95 degrees F. While it is an economical system, it may not feel comfortable to you standing over a floor register.

Given the number of system combinations and localized factors, it would be impossible to direct you in making this decision here. But by considering the questions and suggestions above, you should be able to determine the most economical and satisfactory solution based on your own preferences and locale.

Whichever you choose, the HVAC system will likely be the largest single load on your electrical system. It will normally be set prior to the electrical rough-in since the ductwork needs to be installed prior to that time. Virtually all systems will require 240v service, either for the heating coils, auxiliary/emergency heat (heat pump) or the compressor (heat pump or air conditioner). All systems installed which utilize refrigerant will require that installation be done by an EPA Certified Refrigeration Technician who has been qualified by taking training and passing an extensive exam.

After installation, the system must be evacuated and held under vacuum for a specified time to ensure no leaks before the refrigerant is charged to the system. Any servicing done on the refrigerant side of the system in the future must also be performed by a Certified Technician.

ELECTRICAL CHECKLIST

- Review electrical plan drawings with special attention to
  - Switch locations
  - Receptacle locations
  - High load areas
  - Major appliances
- Determine HVAC requirements and select system to be used; have units set in place and ducting run after the house is framed
- Check total load to house to determine the size of the load centers; make sure non-typical or high load areas for tools and appliances are included
- Install rough-in wiring (after plumbing and HVAC ducts are installed)
- Check the rough-in wiring to make sure all the requirements of the layout plan are met

✗ Make sure all wiring is installed for security systems, intercoms, doorbells, entertainment systems, etc.

✗ Have rough-in wiring inspected and correct any deficiencies

✗ Select all lighting, switch, and receptacle types

✗ After all wall finishing is complete install switches, receptacles, lighting, and make all final connections

✗ Check to make sure all breakers and circuits are clearly labeled in the load centers

✗ Have the final inspection made and the meter installed

CHAPTER 11 CHECKLIST

✗ Select the highest quality kiln-dried framing lumber you can afford

✗ Sill plate attached to foundation

✗ Floor joists attached to the sill plate and reinforced for heavy load areas

✗ Subflooring installed on floor joists

✗ Determine material (wood, metal) and technique (stick built, platform, etc) to be used for wall framing; construct exterior and then interior walls

✗ Install sheathing

✗ Construct roof framing and apply sheathing and roofing; make sure the roof has proper ventilation

✘ Check that proper flashing is used in all areas where something protrudes through the roof and in all valleys on the roof

✘ Install guttering and downspouts

✘ Install exterior wall covering — whether siding, brick, stone, etc.

✘ Complete exterior finishing checklist

✘ Complete plumbing checklist

✘ Complete electrical checklist

12

Kitchens

For most homeowners, the kitchen serves as much more than just a place to prepare and serve meals. In addition to this use, it tends to serve as the hub of family communication. Most people "live" in their kitchen. You may want a place for your children to study as well as a desk where you can pay the bills, read the morning paper, and more. You may want the kitchen to contain both a bar for quick meals and a seating area for serving more elaborate meals.

The style of a kitchen commonly reflects its architectural nature. For example, if you are building a blocky Greek Revival home, your kitchen cabinets may consist of symmetrical panel doors of varying sizes. For an art deco theme, you will probably choose cabinets with clean lines and glass doors or metal accents.

Aside from the architectural style, there are many issues to consider when designing a kitchen.

Size — The kitchen should be large enough to provide sufficient counter and work space. This is easy to say, but not as easy to do, as every cook has different requirements. People who spend more time baking need lots of counter space; those who create gourmet meals on a regular basis will need more storage for their special tools. Some who cook very little will not want to waste square footage on the kitchen. Your kitchen needs to accommodate the features and appliances you want and have enough room to move around, yet do so without requiring extra steps or distance between major workstations.

Purposes of the Kitchen — Besides cooking, the kitchen will probably be used for at least some of your entertainment as well as for eating. Are you comfortable with only a small snack bar, or do you want space for a breakfast nook? Do you need a full-sized table in the kitchen? Will there be multiple hoods, or perhaps an island? These decisions will play into the size of the kitchen. If your kitchen is too small, everyone will feel cramped. If your family is used to congregating in the kitchen, this could be a problem. Better to think it through ahead of time.

Some decisions concerning the kitchen will include:

- **Cabinets** — Base, wall, or corner cabinets? Will the corner ones include lazy susans? Do you prefer open cabinets? Glazed or color? Will you add special features?

- **Appliances** — Appliances take up more space than anything else in the kitchen. Because of this, most homeowners need to select them first to determine space requirements for the layout. At this time, also determine which counter appliances you want to use (mixer, coffeemaker, microwave, etc.) and where they will go.

- **Pantry Space** — In the kitchen or nearby, part of cabinetry?

- **Storage Space** — Pots and pans, canned goods, small appliances, china, flatware, glasses, cooking utensils, etc.

- **Counters** — The type of countertop you choose will have an enormous impact on the look and feel of your kitchen. There are many choices and reasons to consider, including:

- Price

- Weight

- How durable they are

- How much upkeep is required

To save on costs, many homeowners customize each work zone with the type of surface that will be appropriate for use, and put luxury countertops only where they will have an aesthetic impact.

Some of the more common choices are:

Laminate

Laminate is easy to install, and is available in many colors and textures. It is stain and impact resistant. It is also the least expensive countertop, available at $10 per square foot. On the down side, laminate has visible seams that you will not be able to conceal. If scratched or marred, there is almost no way to repair it. Also, you cannot use a knife for cutting if your counters are laminate.

Butcher Block

Made of rock maple, butcher block may cost $16 per square foot. The surface is great for cutting, and develops a certain character from the marring as you use it. It can be almost instantly brought back to new condition by sanding it down a little and adding a coat of mineral oil or beeswax.

The negatives of butcher block countertops are few, but it is quite vulnerable to water. It must be wiped dry, and if used for cutting raw meat or fish it must be cleaned thoroughly.

Ceramic Tile

Ceramic tile is a popular choice because it is easy to install and repair. It comes in nearly any design choice and resists moisture, scratching, heat, and stains. However, the grout used between tiles can become stained or mildewed, and is difficult to clean. Newer epoxy grouts are available which will not stain or mildew. Although they are slightly more expensive, they are well worth the extra cost. Also, ceramic tile can cost up to $80 per square foot, so the budget-conscious homeowner should be careful about choosing tile.

Engineered Stone or Granite

These are the toughest and least porous materials. They are resistant to scratches, stains, heat, and the engineered stone will never need sealing. The cost will be at least $50 per square foot and can run well into the hundreds. These require professional installation and, if damaged, will need professional repair. They are also extremely heavy and this must be taken into consideration if the kitchen is not placed on a slab.

Marble

Marble offers a cool, non-stick surface that is perfect for the serious baker. The cost is fairly low, around $50 to $75 per square foot. However, marble is porous. That means it is easily stained and scratched, and may become discolored. It must also be professionally installed and sealed regularly.

Solid Surface Acrylic

Solid surface countertops are non-porous. That means there will be no germs living anywhere in the counter and it is difficult to stain. Clean-up is easy, and even scratches can easily be buffed out. It comes in dozens of colors and patterns, including some resembling stone. The seams blend easily with edges and integral sinks. However, solid surfacing can be expensive, especially since you will want to choose one of the best brands to get a good quality counter. It requires professional installation, and many homeowners find that solid surfacing is damaged by hot pots or pans.

Concrete

Believe it or not, some contractors who want to create an unusual shape to their countertop prefer concrete. It can be cast right in the kitchen, is heat resistant, and can be tinted nearly any color. Newer types of concrete have less propensity to crack and offer decorative finishing. However, the price of concrete counters is high, unless you are adept at pouring it yourself. If done incorrectly, it can crack. Also, the look can turn out to be somewhat industrial if it is not planned carefully.

Because it is so easy to find a kitchen designer, this chapter will focus on remodeling more than on designing a brand new kitchen.

REMODELING A KITCHEN

Every kitchen design should be planned around the classic kitchen triangle, which describes the way you flow from one area to another during kitchen work. The points of the triangle are:

- The sink
- The stove and oven
- The refrigerator

You should be able to reach each of these three points within the triangle easily (within one or two steps), and there should be nothing blocking you from one point to the next. When demolishing the old kitchen, remember that there may be lead-based paint or asbestos present. These will need special handling and may require special waste stations, because not all trash yards will accept hazardous materials. Be sure to check to see if there is any lead or asbestos present.

If you are remodeling the kitchen, check with the designer and the contractor to see who is going to obtain the necessary permits for electrical, mechanical, or building work. This must be done prior to the beginning of construction.

PLAN THE DESIGN

Even though the kitchen probably has a fairly set design, it will be good to consider changing it. Kitchen remodels hold their value, and if you are planning to sell the house, they can be one of the best choices for updating. The design of a kitchen is really fun. There are only five basic design plans, and if you are remodeling you will probably already have an idea of which one to use. The five basic design plans are:

The One-Wall Design

If your kitchen is long and narrow, it may have been built in a linear fashion. This is a design that allows for unimpeded traffic flow, but if there is enough square footage, consider moving one point of the triangle to the opposite wall.

The Corridor Design

Mostly seen in apartment living, the corridor plan has a triangle but is somewhat crowded. The refrigerator may be found at one end, with the range and sink directly across from each other — with very little room between. If your kitchen space is at a premium, this may be the most affordable way to design the kitchen.

The L-Shape

An L-shaped kitchen arrangement features a refrigerator on one end of the L and the range on the other, with the sink located near the center on one leg or the other. It gives more counter and cabinet space, but the corner counter may be hard to reach.

Double-L

A double-L layout contains a small workstation, often with a cooktop and a second sink, within the larger workstation, which is used for food prep. There are plenty of open counter spaces, though no more cabinets than the typical L-shaped kitchen.

The U-Shape

A U-shaped design gives the cook good work flow and plenty of cabinet space. Designed like the corridor, but with the end closed, there is room for the range or even a sink on the short end.

The kitchen remodel may require that you add joists to strengthen the floor to support heavy appliances, granite countertops, or a kitchen island. The new kitchen may have an automatic ice-maker, an island with a sink, or you may have chosen to move the main sink. At any rate, you may need to rough in new water supply lines and drainage pipes. If you are doing this yourself, it may be slow going; consider hiring a professional plumber.

The electrical service should be at least a 200 ampere to support the kitchen. An electrician will probably want to run new wires for lighting and appliances. Code will require that the wiring meet current standards. Have an inspector approve the electrical system before hanging drywall.

You will need to store your new materials somewhere until it is time to install them. Find out whether the designer or a subcontractor has a warehouse where products are normally delivered. If you are interested in sending old cabinets and used fixtures somewhere other than the local landfill, consider using them as extras in the laundry or garage area. You can also donate such products to local salvage centers, or even to non-profits that can use them for rehab projects.

Use clean, sturdy boxes to pack up the items that will not be used during renovation. Store these away from your project to protect them from damage. This will include pictures, mirrors, furniture and other items that are in the kitchen or in adjacent rooms. The vibration from the construction can damage them if they are not removed. Before renovation begins, pack up the contents of your kitchen. Start with the items you use the least.

When the day for demolition arrives, pack the most used items, which you'll want to store in the most accessible containers. Also, dust is one of the biggest problems when revamping a kitchen. Be sure to cover the floors and hang a dust curtain over doorways. Let workmen know which ways you want them to enter and exit the work space, and be sure they always use this path. You can ensure this by hanging plastic sheeting, taping it shut over other doorways. Either tape over the HVAC vents or do not let the central heat run while the workers are sanding; it will blow the dust throughout the ductwork and you will have dust forever!

CHAPTER 12 CHECKLIST

✖ Budget   ✖ Design and Layout

✖ Demolition (if remodel)   ✖ Rough Framing

✖ Electrical   ✖ Major Appliances

✖ Wall Outlets   ✖ GFCI (for all outlets)

✖ Lighting   ✖ Main

- ⚒ Task Lighting
- ⚒ Accent
- ⚒ Gas Service
- ⚒ Plumbing
- ⚒ Sink
- ⚒ Faucets
- ⚒ Dishwasher
- ⚒ Icemaker
- ⚒ Filters
- ⚒ Heating & Air Conditioning
- ⚒ Fans and Ventilation
- ⚒ Insulation
- ⚒ Drywall
- ⚒ Windows
- ⚒ Doors
- ⚒ Finish/Stain
- ⚒ Cabinet Selection
- ⚒ Base
- ⚒ Wall
- ⚒ Tower/Storage
- ⚒ Open Shelving
- ⚒ Island Base and Overhead
- ⚒ Valances
- ⚒ Crown Moldings
- ⚒ Walls
- ⚒ Baseboards
- ⚒ Countertops
- ⚒ Sink
- ⚒ Backsplash/Tile
- ⚒ Appliances
- ⚒ Garbage Disposal
- ⚒ Microwave
- ⚒ Refrigerator
- ⚒ Cooktop
- ⚒ Oven
- ⚒ Freezer
- ⚒ Dishwasher
- ⚒ Other

- Flooring
- Covering
- Woodwork
- Structure
- Paint Walls
- Wall Covering
- Number of Circuits and Electrical Load

13

Bathrooms

BATHROOM SIZES

When looking through glossy home magazines, who does not pause and dream when looking at a centerfold photo of a luxurious master bath? In the magazines, these rooms sure look terribly desirable, but if you stop to calculate the square footage they consume and the cost of filling that square footage with tile and fixtures you may quickly reconsider. Unless you are captivated with the idea of bathing in a gymnasium, you can probably do with a much smaller space.

Much of this advice is subjective; if all the houses in your neighborhood have huge bathrooms with sumptuous fittings, then you will probably want to remodel yours in this manner. In some areas, it is practically a requirement. In others, it is a luxury.

Many of the people reading this book are remodeling a master bathroom, trying to make use of the available space or, for new construction, trying to make the best use of the allotted space. Instead of laying out a hanger-size bathroom and outfitting it with every available gadget in the home magazines, you might be better off by utilizing some of the space to increase closet size (maybe a walk-in), adding storage, or increasing the usable size of the bedroom. This could also give you more available budget to fill the space you do use for the bathroom with better tile, fixtures, and flooring.

A toilet, sink and shower can fit into a space as small as 5'x8'. If you need a double sink and a shower/tub combination, the minimum will stretch to about 8'x9'. Code requirements define the minimum amount of space for each fixture in the bathroom, but adding just a few inches here and there can make the difference between feeling cramped or comfortable and spacious. Most codes require 21" in front of sinks and toilets, but 24" is much better, and 30" will feel very comfortable. Remember that these clear spaces can overlap, as you are not likely to be using the toilet and the sink at the same time. Having more than 36" around a fixture is a waste of space, which can be put to better use elsewhere.

To figure out how much space to allocate for the bathroom, sit down with your partner and ask yourselves a few questions. Do you have similar schedules, or does one of you go to bed earlier and get up earlier? A dressing area lets you turn on the lights, get dressed or undressed, or find clothes without disturbing the other. Do you both need the sink and the mirror at the same time? A double vanity takes a lot of room, but may save a marriage. If you always shower you may not need a tub in the master bath, but if you enjoy a good soak a tub is a requirement. What about the seemingly obligatory Jacuzzi? How often would you use it? They take up a lot of valuable real estate, put a large dent in the budget, and often become merely a bathroom ornament. Your bathroom needs to be sized for your lifestyle and the way you will use it.

SIMPLE OR ELABORATE

The selection of components for finishing baths is almost unlimited, and prices range from economical to extremely expensive. As you look through hardware superstores, supply house displays, and catalogs, consider usability, practicality, durability, appearance, initial cost, and the ever-important resale value. As with a kitchen remodel, the cost of remodeling a bath can be recoverable or add value if you make choices wisely. But just as overbuilding the neighborhood with a new house or addition will result in not being able to recover your costs, you can outspend the ability to recoup the cost of some bath components. As you are preselecting components during the budgeting process, these things should be a part of your decision:

- **Cabinets and Lavatory** — Your cabinet and lavatory layout will be determined in part by the space available after the fixtures are laid out in the available space. Cabinets come in many styles and lavatories can range from a freestanding basin or wall-hung sink to a one-piece top on a cabinet (see countertops). For cabinets, consider the storage requirements you need for grooming accessories such as curling irons and hair dryers, toothbrushes, razors, cosmetics and lotions, medicines, towels and paper products, and so on. Having to leave the bathroom for something you need to get ready with every day will be a very unrewarding experience. The selection of a cabinet finish will give you an opportunity to coordinate the colors and textures in the room along with the tile, countertops, and flooring. Cabinets come in a wide price range and offer an opportunity to find a good price.

- **Mirrors** — Mirrors are an important component of the bath. They need to be adequate for applying make-up, fixing hair, and other personal grooming tasks. A large mirror will reflect the light and also give the room a larger feel. Mirrors are available as a plain plate glass mirror in any size you want (from a glass shop) and offer a low cost solution with a good look. They are also available in combination with a built-in medicine cabinet and with side wings, which fold out and allow a view of the back of your head. They come with and without lighting, framed, and beveled or simple-seamed edge. Here again, the choices are seemingly endless and will be determined by your style and budget.

✗ **Lighting** — Most people do not complain that there is too much light in the bathroom. This may need to be the best lit room in the house. Consideration needs to be given in light placement so that there is ample light to use the mirror without creating a glare. General lighting should be provided throughout the bathroom, with task lighting provided at every functional area. Often, decorative lights look great in the room but provide inadequate lighting where it is needed. Be sure to check the allowable wattage of the fixtures you choose.

✗ **Ventilation** — Moisture is always a problem in a bathroom, particularly if there is a shower. Unless provision is made to remove it there will be problems with mold, mildew, and peeling wallpaper and paint. A ceiling fan which is ducted to the outside is preferred, as it completely removes the humid air from the house. A vent which discharges into the attic can easily transfer the problem from the bath to the attic. In cold weather there will undoubtedly be some condensation, which will cause problems later. Fans are available as a stand-alone or combination fixture (includes a light and/or heater). This is a good opportunity to get additional lighting in the bath at a low incremental cost.

✗ **Countertop/ Sink** — Countertops and sinks are available in many styles, materials, textures, and colors. Some of the more commonly used are:

- **Laminate/Drop-in** — You can purchase sections of rolled edge laminate covered countertops, which can be cut and installed by the average do-it-yourselfer. A hole is cut and a drop-in sink is installed (also available in a variety of styles and materials). This is probably the low cost choice. Another style has a squared edge. This style can be built by constructing the plywood base in place, gluing the laminate on to the base, and trimming the edges with a special router bit.

- **Stone** — A currently popular material for use in countertops is stone, most often marble or granite. For many applications, they can be made in one piece for a seamless finish and can be purchased with a shaped edge in several styles. Granite is the most popular choice for countertops because it is much harder than marble and has greater stain, heat, and water resistance. It is also more resistant to abrasion. The price varies based on the complexity of the edge desired. Drop-in and bottom mount sinks can be used.

- **Engineered stone** — Engineered stone is a quartz-composite product mixed with colored pebbles, polymers, and epoxy. It has an even pattern and more color options than natural stone. It is an extremely durable product that takes heat and resists stains. Engineered stone is installed using epoxy and is about the same cost as granite. Drop-in and bottom mount sinks can be used.

- **Tile** — As seems to be the case with most other bath options, tile comes in a seemingly endless variety of sizes, colors, textures, and patterns. Ceramic tile offers a smooth, glossy, and easy-to-care for finish. It is probably the most popular choice for walls and backsplashes, but is also used for counters. Clay tiles have a matte finish and are available in many colors and textures. They are equally waterproof and stain resistant. Epoxy grouts, which will not stain, are now available for use on countertops and are recommended. Drop-in sinks are used with tile.

- **Solid Surface** — Solid surface countertops offer seamless acrylic faces with seams that are only visible from the underside. With solid surfacing, sink and counter materials can be integrated to create a fluid, graceful line. Solid surfaces come in a rainbow of colors, patterns, and styles, including stone and glass look-alikes. They are stain and heat resistant, with more edging and border options than natural stone. Scratches are easily softened with a non-abrasive scrubbing pad.

✕ **Plumbing Fixtures** (See Plumbing Section in Chapter 11)

✕ **Bathtub/Shower** — There are four basic choices when consider a bathing unit:

- **Combination Tub/Shower Unit** — a fiberglass/acrylic combined unit that includes the tub and wall sections. They come in one-piece units for new construction or with separate wall panels for remodeling use. Most have molded-in soap dishes and handles and are available in different colors. These are easy and relatively inexpensive to install.

- **Bathtub with Wall Treatment** — essentially the same as the combination unit, except that instead of having the wall section as a part of the package, the tub is installed and then a wall treatment such as tile is used. This gives an expanded choice in tub selection and material (enameled cast iron for instance) and a much greater opportunity to coordinate with tile selections for the tub surround.

- **Separate Stand Alone Shower and/or Tub** — If you have the space and budget, a popular option is separate units. You can have a tiled walk-in shower and a separate tub.

- **Jacuzzi** — As mentioned above, a Jacuzzi is a high-cost option, from the initial purchase cost to the installation and operation.

Before adding one of these to your bath, be sure you will use it enough to justify the expense.

⚒ **Flooring** — There are many choices available for flooring in a bath. The main choices are wood, tile, and carpet.

⚒ **Wood** — Available in a variety of woods, colors, and finishes, wood flooring is a popular choice for many houses. It is easy to clean but is subject to water damage.

⚒ **Tile** — Also available in many colors and textures, tile is easily cleaned but the grout should be sealed to prevent stains from dropped make-up and other accidents. Tile is waterproof and can survive spills.

⚒ **Carpet** — Carpet is subject to water damage, mold, and mildew. While it is not unusual to see it used in baths, thought should be given to the maintenance requirements.

⚒ **Electrical Outlets** — Just as there is never too much light, there are never too many outlets. Number and location may be addressed by the local code, and circuits must be GFCI protected. Determine how many appliances will have to be plugged in at once (for example, curling irons, hair dryers, radio, and phone charger) when determining how many you need. Codes do not allow outlets to be placed within a specified distance of bathtubs and showers.

TILE WORK

Just like choosing the right accessories to match an outfit, the right tile choice can influence the impact of your bathroom. Whether you are using colored ceramic, a clay tile or a pattern, tile can bring texture and enhance the overall look of the bathroom.

Tile Selection

One of the most popular wall treatments in bathrooms, particularly on surfaces which will be exposed to water, is tile. Tile is a durable surface, and can be installed by many do-it-yourselfers with a little training. It is relatively easy to keep clean and comes in a myriad of colors, styles, and designs. Before you go to the tile dealer to select your tile, make some preliminary decisions about what you want. You can choose from glazed ceramic tiles, matte-finished clay tile, and stone in either plain or textured surfaces. It will help the decision process if you can narrow the scope of what you want somewhat before you go in. However, be warned that no amount or pre-visit planning will prepare you for the thousands of available patterns, colors, and textures available at most tile and carpet dealers. You will undoubtedly find many tiles to make a selection from. Ask the dealer questions about ease of cleaning, durability, tendency to spot or discolor, and care requirements before making your final decision. Take samples of any wallcoverings you have in mind to make sure all the materials blend or match as you expected. Most dealers will allow you to check out samples and take them home to view in the actual room (for a remodel) so you can see it against the fixtures.

Design

Tile designs are nearly as limitless as the selection available. You use just the tile you picked, add a border, intersperse tiles with pattern or design, alternate colors, and on and on. Most of the options add a lot to the interest and not very much to the cost. The tile dealer will typically have a lot of literature and suggestions so take advantage of their experience.

Installation

Many of the hardware superstores offer free classes on installation which, with a few special tools, can produce professional looking results. Tile cutting saws can be rented and others are reasonably inexpensive. If you are doing a remodeling project that involves removing old tile, a few words of caution are in order. The primary way of removing the old tile from the wall is with a hammer. Broken tiles are extremely sharp — in the razor category. Small chips fly around so good eye protection, like goggles that seal all around, are absolutely essential. This cannot be over-emphasized. In addition, bathtub fixtures and the tub itself must be protected from damage from falling tile pieces.

For tubs and showers, a cement backing board should be put on the wall and the tile installed on the board. This product is waterproof, will not swell and crack, and provides a stable surface with good adherence properties for the tile. It is normally attached to the stud wall with galvanized screws. Careful layout of the tiles is a must for a do-it-yourself job. If you end up with partial tiles to finish a row (almost a certainty) you want equal sized pieces on each end. Therefore, you must start in the middle. This is one of the things you will learn in the class. Instruction is also available on video and several Web sites, but there is no substitute for hands-on work.

CHAPTER 13 CHECKLIST

Fixtures

Bathtub/Shower

- ✗ Large enough to accommodate each user

- ✗ Non-slip bottom surface or floor

- ✗ Faucets in comfortable reach and properly oriented (hot on left)

- ✗ Soap, shampoo storage, and towel bar conveniently located

- ✗ Shower head at a comfortable height for all users

- ✗ Easy, safe entry and exit

✖ Grab bar in the tub or shower

✖ Sufficient clear space in front of tub/shower

✖ Glass shower doors use safety glazing

✖ Shower door swings into bathroom, not into shower

Toilet

✖ Located a minimum of 16" from center line to wall or cabinet

✖ Sufficient clear floorspace in front

✖ Doors do not open into toilet space

✖ Paper holder comfortably located

✖ Toilet design allows ease of cleaning

✖ Lavatory

✖ Sink height comfortable for all users

✖ Distance from center of sink to wall 15" minimum

✖ Minimum of 30" clear space in front

✖ Door does not swing toward cabinet and block access

✖ Backsplash or tile is waterproof and sealed to lavatory

Walls

✖ Walls that come in direct contact with water have a waterproof covering (tile, fiberglass, stone), sealed to prevent unseen leaks and for ease of cleaning; this includes enclosures around tubs, showers and behind sinks

- Walls subject to splashing have water-resistant covering (wallpaper or enamel paint)

Floors

- Finished in a way that resists water damage
- Easy to clean
- Non-slip surface

Mechanical/Electrical

- An efficient ventilation system
- Bathroom will be comfortably warm
- Sufficient electrical outlets, all with GFCI protection, located away from the tub/shower
- Adequate lighting for all bathroom activities
- Shutoff valves on all water lines to sink and toilet

Storage

- Adequate shelf or counter space around the lavatory
- Sufficient space to store toiletries, grooming equipment
- A space for towel storage in or near the bathroom
- Medicine and cleaning supply storage out of the reach of small children

14

Putting on the Finishing Touches

DRYWALL

Installing drywall, while somewhat strenuous, is not a difficult job. So long as you have a helper for lifting the drywall into place, the actual attachment is relatively easy. Finishing the drywall will take a little practice to become proficient.

Purchasing Drywall

Determining the number of sheets of drywall you need is relatively easy. Just determine the number of square feet of wall and ceiling you are going to cover and buy that much drywall, allowing for scrap. You can consult the dealer for help in determining the rolls of tape, buckets of drywall compound, and number of nails or screws you will need. If you have outside corners, each one will require a metal cornerbead.

Hanging Drywall

Drywall comes in 4' wide sheets (normally 8' long) but is also available in 10' and 12' lengths. It is normally oriented with the long dimension vertical (on the walls). The objective is to minimize the length of joints to finish, so it is perfectly okay to hang it horizontally if that accomplishes the goal. Drywall of 1/2" thickness is usually used for walls and 5/8" thick is used for ceilings. Carefully lay out your drywall sheets to minimize joints. Measure and cut the drywall sheets with a sharp utility knife. After you have scored it on one side you can lay it on a piece of 2"x4" wood and snap it along the cut. Then turn it over and trim the paper on the other side.

Begin by installing the drywall on the ceiling. Before starting on the ceiling, you will need to construct a pair of "T" braces. You can make these with a 3' long piece of 2"x4" wood (2x4), nailed at 90 degrees to the end of the 2x4, about an inch longer that the distance from the floor to the ceiling. These braces will be used to help lift the drywall for the ceiling into place and wedged to hold it in place while it is being fastened. Before raising the sheets into place, be sure to cut out required openings for lighting, fans, and other necessary items. Fastening is done by either nailing or screwing into the ceiling joists; using screws is the most common practice at this time. If you use a nail, be sure to use one designed for drywall and hammer it below the surface of the drywall so that the head of the hammer leaves a dimple in the surface. If using screws, the screw heads should also be below the surface. These dimples will be filled during the finishing process. Fasteners should be spaced at 6" intervals starting at the center of the sheet. Using a chalk line will help you in getting the nails or screws in the center of the joists.

Once the ceiling is done you can start on the walls. Before hanging drywall on the wall, carefully lay out and remove material where windows, doors, switches, and receptacles are located. Once the holes are marked you can start the hole with a drill and finish it with a keyhole saw. Butt the wall panels tightly against the ceiling. Wall fasteners begin 4" from the ceiling and are spaced 6" apart. When you have an outside corner, measure and cut a metal cornerbead and fasten it in place. You do not need to dimple the fasteners on the cornerbead because they will be covered with plaster.

Joint Taping

The joints where pieces of drywall meet are finished using a process called taping. To have a smooth surface, the joints must be filled and finished so they appear to be flush with the rest of the wall. To facilitate this, the edges of the drywall are manufactured tapered. To reinforce and smooth the joint, a layer of joint compound is first spread into the recess created by the tapered edges using a 5" taping knife. Smooth the compound until it is flush with the rest of the sheet. Then center the drywall joint tape over the joint and press it into the compound. The compound will squeeze out through the holes in the tape and along the edges so make sure there is still some compound under the tape. Once the tape is firmly in place, smooth out the compound with your taping knife. After you do this, fill all the fastening dimples with compound and smooth with the taping knife. Also apply compound to the cornerbeading, using the 5" knife to smooth the compound from the edge of the bead, over the fasteners, and out onto the board. Allow all the compound to dry completely, usually for at least 24 hours. It is important not to apply too thick a coating at one time since the compound will shrink and crack.

After the first coat of compound dries, you will need to apply two more thin coats. The first should extend a few inches beyond the first coat. After it dries, the third coat should extend about 6" to each side of the joint and should be done with a 10" blade. After the third coat is thoroughly dry, sand and smooth all surfaces lightly with medium grit sandpaper.

The inner corners and the joints between the wall and ceiling must also be taped. Tape is available that is pre-scored down the center and folds easily to facilitate this. The procedure is the same as for wall joints except that more care must be taken in the corners not to cut the paper with the edge of the joint knife. Sanding is also required after the third application of compound.

WARNING!! If you are doing this as a remodeling project, be forewarned that sanding dried joint compound is akin to tossing up a box of face powder and letting it hit the floor. It goes everywhere and gets into everything. You would do well to tape plastic over the doors leading anywhere in the house and to tape over any HVAC ducts. You should also wear a dust mask while doing the sanding and clean-up. An exhaust fan in a window to give the room a slight negative pressure is also a good idea.

PAINTING AND FINISHING

Painting is one of the primary areas where people tend to think they can save money by doing it themselves. While this is probably true, you also have to consider time as having value. A paint crew can come in and paint an entire house, including woodwork, in two to three days. The do-it-yourselfer may require two weeks to complete the job. This can seriously delay your schedule and cause you to lose subs to other jobs while they are waiting for you to finish. You have to make this decision for yourself, but be sure to consider the real cost of saving the money.

The array of wall finishes, from paint to faux finishes, stone to wallpaper, rough-cut cedar to fine mahogany, are endless. Your choices are limited only by your sense of design and your budget. The most common wall finish is paint so that is primarily what is discussed here. For specialty materials, instructions for their application are available where you buy the product. There are a few finishes the average do-it-yourselfer may not be able to accomplish. Many people feel that wallpaper, for instance, is beyond the ability of the average person while, in fact, putting up modern pre-pasted papers requires only a little on-the-job experience, a good straight edge, paper tray, razor knife, and sponge. In most cases, all applications begin with a sealer.

For drywall, it is recommended that the compound be allowed to dry for five days before painting. The first coat that is put on a new drywall surface is a sealer coat. After the sealer, sand very lightly with fine grit sandpaper. The next coat is a primer coat. Primer is available with a mildew repellent included and should be considered for use in damp areas or anywhere mildew or mold could be a problem. You can have the primer tinted if you are planning to have a dark colored paint finish coat to reduce the number of coats required, but beware: Primer does not take the tint like finish wall paint. It will not be the color you expect (for instance, my dark chocolate brown tint turned out mauve in the primer) but may be close enough to let one top coat do the job.

Again, lightly sand after the primer coat with fine grit sandpaper. Remember to wipe down the walls with a tack cloth after each sanding to remove the dust and particles. If you are papering, the wall is now ready for the paper. If you are painting, the final coat will be wall paint in the color of your choice. Consider the use of the room when selecting paint type. Most often people use flat wall paint with semi-gloss enamel in painted woodwork for just a little sheen. This combination works well in a living room or bedroom where the walls are not subject to a lot of dirt and moisture, and will thus survive with a little sponging now and then. For kitchens and bathrooms, you might consider using semi-gloss enamel on the walls and possibly ceilings. This is better for resisting moisture and will withstand some hard scrubbing to remove grease and food splatters.

CEILINGS

Ceiling finishes are typically somewhat more limited than walls. Assuming that you are not planning to replicate the Sistine Chapel, your choices are limited to either paint or one of the modern textured finishes. For a painted ceiling, the surface with all its joints must be carefully taped and finished.

Unlike walls, where in the finished room only small sections are seen at one time due to wall hangings, pictures, furniture, windows, and doors breaking the surface, the ceiling is nearly all visible at once. In addition, it is usually seen in a raking light (light shining from the side) which shows every hill and valley in the surface. Textured ceilings are frequently used for this reason. If you do plan to paint, the same procedure applies to the ceiling that was used for the walls.

If you are going to use a textured finish, perfect flatness is less of a concern. There are several textured finishes to choose from; probably the easiest to apply is the spray-on finish. This is a thinned-down joint compound mix which has small irregular shaped particles that give the texture its nubby look. There are also special texturing compounds made just for this purpose. This is a fairly easy application for the do-it-yourselfer using rental equipment. You should do this type of ceiling before painting the walls and should tape the edge and tape plastic sheeting over the walls to prevent spray from getting on them. Follow all safety instructions that come with the spray gun and material, especially wearing goggles and a breathing mask. Another texture is made by applying joint compound on the ceiling with a trowel and then pressing a sponge into the material to form a pattern. One word of caution is in order: A light sponge yields a smoother surface with a more professional and finished look. You can also create a stipple finish or a swirl pattern using a stiff brush on the applied joint compound.

WOODWORK AND RAILINGS

Woodwork and interior railings can either be stained or painted. Painting can give a more formal finished look and allows you to use a less decorative (i.e. less expensive) wood than staining. Whichever method you choose, you will save time and frustration by painting or staining the woodwork before attaching it to the wall, then filling the nail holes and touching up with paint.

GARAGE DOORS

There are many styles and finishes of garage doors to choose from. If the garage is not going to be heated, you can save some money by getting uninsulated doors. Doors are available in several styles to match your house style. They are available in wood and galvanized steel panels. Most are pre-finished and some are available in a variety of colors. They also come with a variety of windows starting from none. The size of your garage will determine whether you have a single, large double-door, or two singles (for a two-car garage). While a single, large door will save you money, at least on installation, remember that a single double-door will probably result in the cars being closer together. Double doors are typically 16' wide. Singles come in 8' and 9' widths. The 9' width will allow a lot more comfort in putting a large car or truck in the garage. Door installation can be done by the do-it-yourselfer, but if you selected one with a large spring mounted on the wall above the door which has to be wound up, you would do better to have it installed by an expert. Winding the spring is a dangerous operation, and many people are injured each year trying to do it without training.

POOLS AND EQUIPMENT

The installation of a pool is a major consideration from a financial standpoint. Not only is the installation expensive and upkeep costly, sometimes having a pool is actually a liability on resale in many areas of the country. However, for many people a pool is an integral part of their lifestyle and provides immeasurable recreational value for the family. In some areas they are considered a necessity. Pools come in many sizes and styles including freestanding above-ground pools, in-ground pools with liners, in-ground fiberglass pools, and in-ground concrete pools. Planning for the building of a pool is like any other construction project. You need to check on codes and restrictions, make a budget and schedule, look at materials and designs, select the subcontractors, get the financing, and complete the project. The pool design and installation will most likely be done by a contractor who specializes in this area. They can recommend the sizes and specifications for pumps, filters, cleaning equipment, etc. They can also give you information on water testing, treatment, and maintenance requirements.

DECKS

If you are planning a deck, the design and configuration have undoubtedly been done as a part of the design for the new house. As a remodeling project, adding a deck is a popular option, or you might be replacing an existing deck which has deteriorated and is no longer attractive (or safe). Once you have the design for the deck, the biggest question is what material to use. The basic choices are natural wood and composites. Natural woods include:

- Cedar

- Redwood

- Mahogany

- South American hardwoods

- Salt treated pine (pressurized lumber)

Availability and cost of these options may vary in different parts of the country, but the salt treated lumber has become the low cost standard in most areas. The choice depends on the look you want, what is available, and the cost.

The composite deck materials are the new entrants in this market. They have a lot of green power since they are made of recycled materials. They are available in a variety of colors. Though many offer impressive warranties, most composite decks have not been in place long enough to see if they are as truly low-maintenance and durable as the manufacturer claims. In addition, the composite boards are not as strong as natural wood and therefore require more support (support joists placed closer together). Since the supports are natural wood, the true, maximum life of the deck is the life of the support wood. Decks are not normally painted, but they can be. You should not plan to use non-pressurized framing material and count on the paint to prevent weather damage. Get advice from your local paint supplier concerning how to best paint the wood you are using for your area.

CHAPTER 14 CHECKLIST

- Hang drywall, cutting all required openings

- Tape the joints and finish with joint compound and final sanding

- Seal, prime, and paint drywall

- Install wallpaper, paneling, other wall coverings

- Paint or stain all woodwork and railings, then install and touch up paint

- Select garage doors and install

- Plan for pool design and installation as a separate construction or remodeling project

- Complete design and planning for decks

- Make decision on type of materials to use

- Construct decks

15

Potential Dangers & How to Prevent Them

RADON, ASBESTOS, AND LEAD

These three environmental and health dangers were all discussed in Chapter 8 under Environmental Testing. For new construction, the EPA has recommended guidelines for the prevention of potential radon problems and ease of installing radon control systems. The recommendations vary depending on the Zone you are in, relative to the risk of high radon levels indoors. You can look over the EPA Zone map at **http://www.epa.gov/radon/zonemap.html**. Check on the Building Codes for your area to see if preventive measures are required. If you are in Zone 1 you may want to take preventive measures, even if codes do not require it, to reduce the cost of eliminating the high levels later.

MOLD

You are much more likely to be faced with mold in the renovation of an old house than with building a new one. Mold is any of several types of fungi. Molds are present in indoor and outdoor air, and the particles or spores are much too small to see with the eye. They can grow on any number of materials in the presence of moisture, including carpet, wood, paper, food, and even insulation. The spores are like seeds and travel through air looking for a place to grow. If the house you are renovating has had roof leaks, damp basements or crawl spaces, plumbing leaks, inadequate venting in bathrooms, or dryers not being vented to the outside, there is a likelihood that mold exists. Part of the problem with mold is that the material on which it grows may be damaged. The larger problem, however, is the health hazard it presents. The spores or tiny fragments are inhaled by the people living in the house. Depending on the exposure levels, time exposed, sensitivity to mold, age, and other factors, mold can aggravate and worsen asthma and lead to upper respiratory conditions such as:

- Nasal and sinus congestion
- Coughing, wheezing, and breathing problems
- Sore throats
- Skin and eye irritation
- Upper respiratory infections

If mold is suspected, you can inspect the basement for water damage or signs of excess moisture and other areas for water stains on walls, ceilings, and woodwork. The best tools to use for mold are your nose and eyes. If you see mold or if you can smell an earthy, moldy smell, assume it is there. Testing for mold is considered by many experts to be a waste of money; money which could be better spent eliminating the problem.

ELIMINATING MOLD IS A SIX-STEP PROCESS:

1. **Identify the moisture source and eliminate it** — Several possible sources have already been mentioned. Also, check for good ventilation in basements and especially crawl spaces.

2. **Dry all the wet areas and materials** — You may need to move items around to allow for better air flow. Use fans and dehumidifiers as necessary. Move wet items away from walls and off floors.

3. **Remove contaminated materials and dispose of properly** — You should consider contracting Steps 3 and 4 to a firm which specializes in mold abatement. Specialized clothing, respirators, and other protective equipment should be worn when in contact with mold laden materials. Care must be taken to prevent the dislodging of mold particles and spreading them to other areas, such as hanging plastic sheeting and bagging of disposed materials.

4. **Clean surfaces** — Again, you should consider contracting this work as it also requires the use of specialized equipment. Mold growing on non-porous surfaces can usually be cleaned but safety considerations are of paramount importance. Check with your State Environmental Services Department for local requirements and recommendations.

5. **Disinfect contaminated surfaces** — If you wish, you can spray a mild bleach solution (1/2 cup bleach to 1 gallon of water) onto the areas where mold was removed to kill any mold missed by the cleaning. Wear skin protection and goggles to protect your eyes during the spraying. Get air moving across the area to prevent breathing in the bleach fumes. Clean up any of the solution that runs off the area being cleaned, but do not wipe the area you are disinfecting.

6. **Keep a watch on the area to catch any signs of additional moisture** — This can lead to new mold growth. If moisture appears, repair the problem immediately.

For new construction, the concern will be preventing mold in the future. The techniques for doing this include carefully installing vapor barriers under concrete slabs and in crawl spaces, making sure water drains away from the house, care with installing flashing, and careful checkout of the plumbing. Also, insulate any lines and ducting which may sweat.

TERMITES

Most local Building Codes will require treatment for termites as a part of the project if there is any new foundation work. This is normally done by having a licensed termite control company spray a heavy chemical barrier in the foundation trench. So long as this barrier is not disturbed, the company will guarantee protection from termites for a predetermined time period. For remodeling, if you expose old stud walls or are disturbing dirt around the foundation, it is a good idea to get a termite inspection. These are relatively inexpensive and can prevent unpleasant future surprises.

CARBON MONOXIDE TESTING

Carbon monoxide (CO) is a colorless, odorless, poisonous gas. It is produced by the incomplete burning of solid, liquid, and gaseous fuels. Appliances fueled with natural gas, liquefied petroleum (LP gas), oil, kerosene, coal, or wood may produce CO. Burning charcoal produces CO. Running cars produce CO. The symptoms of CO poisoning include headache, shortness of breath, dizziness, fatigue, and nausea. If you have appliances, heating, fireplaces, or portable heaters fired with any of the above, you should install CO detectors in your house. They should be in the room in which the appliances operate and in sleeping areas. CO detectors are about the same size and cost of a smoke detector, and actually look similar to one. The alarm will go off before the CO reaches the danger level, giving you time to take action.

CHAPTER 15 CHECKLIST

- If you are remodeling, check your home for radon; mail-in test kits are available at most hardware superstores
- If you are planning new construction and live in a radon high probability zone, check Building Codes for required or recommended preventive measures
- Test for asbestos and either seal it or have it removed
- Test older homes for lead-based paints and seal or remove it
- Check for mold in existing houses and take steps to eliminate moisture sources and existing mold colonies
- Install vapor barriers and seal possible areas for leaks in new construction
- Get a termite inspection if not current in existing houses, pre-treat ground for new construction
- Install CO detectors if using fuels likely to generate CO

16

Potential Problems & Common Mistakes

The main reason you probably are reading this book is so you can be your own contractor and save money. The reality is that owner-builders, not being part of the industry, can actually pay more for materials than builders. Plus, you will have to buy or rent tools that the pros already have. Of course, it is entirely possible to save money — and lots of it. Not every aspect of the project can be DIY. Knowing when to hire it out and when you can step in and do work will go a long way toward saving money; plus there are other small steps you can take to save.

The building process is complex, and unless the contractor (that is you!) has a great deal of knowledge about the construction business, mistakes can be costly. Many owner-builders lack technical knowledge and end up relying too much on suppliers or subcontractors to give them information. For example, some even purchase the wrong kind of material for the job, and then will end up paying twice for materials due to lack of leverage with the supplier.

Renovation projects that look so easy on TV often turn into nightmares. With older homes, it is almost a given that each time you remove a wall or floor, you uncover something else that needs repair or replacement. There are also legal liabilities to consider.

Building or remodeling your own home can bring a deep sense of satisfaction and save you tons of money. Everyone should try it at least once, as long as the answer to two questions is yes:

1. **Do you have the time?**

2. **Do you have the money?**

I hope that every reader will check their final design plan against this chapter. Avoiding the most common mistakes is at least a good start.

DESIGNING A HOUSE THAT IS TOO UNIQUE

The phrase "custom home" generally means that one of the main reasons you are choosing to act as your own contractor is because you want a house that is different from everyone else's. But assuming that you can build everything exactly as you would like it often turns out to be the biggest mistake. This is because if your taste differs too much from the norm, you will be stuck with a property that you cannot resell. This is probably okay if you do not need the money, but for most homeowners the value of their home is a significant part of their investment portfolio.

I have seen a custom home with turquoise concrete and one with river rock and concrete walls both languished on the market. I watched an owner-builder enter an old, elegant house complete with a huge ballroom and chop it up into so many rooms that within a year, he was financially drained, unable to sell, and filed bankruptcy.

Your own mistake may not be this drastic; building a house that is a great deal bigger or smaller than the others in the neighborhood can have nearly as bad an effect. Take care to look at what exists around your property.

If the neighborhood is not established yet, check with the Homeowners' Association to see what is being built. They will probably also have minimum and maximum square footage requirements — stay within those.

FINANCING

There are several ways that you can make financing mistakes when building a home. One of these is applying for the construction loan too early. Although in Chapter 4 we discussed learning all that you can about lenders far in advance, applying for a loan too early can be a waste of time. Rates change, loan packages change, and you may end up disclosing unnecessary information — which can ultimately hurt your chances of getting a loan.

In addition to the information end of things, you can waste money by applying too early. If you are not ready to submit the plans to your local building agency, the lender will not fund the loan, because he must wait on permits to be issued before he makes the funds available. In the meantime, you have put in an application, and your credit report and appraisal are going to expire. This is okay, because you are able to renew them, but it will cost some extra money. So it is best to apply for the loan about the same time you submit final plans to the building department — around 60 days before you plan to break ground.

Submitting Multiple Applications

When you make an application to a bank or a broker, they submit the loans to lenders. There are only about 10 national lenders offering all-in-one loans (construction plus permanent). So your loan can be submitted to the same lender via two different bankers you have selected. If this lender sees any discrepancy in the information you have submitted, you may be turned down — and if he was the only lender who was perfect for your needs, you could end up without funding!

To prevent this, ask your loan officer who will fund the construction loan. Be sure that you are clear on the loan process, and stay on top of its progress. If there are problems, try to resolve them. And if the problems do not get resolved, do not feel obligated — just move on to the next loan officer.

Not Starting Out with Enough Funds to Finish

There is nothing worse than running out of money while building a home. Plus, it is impossible to know exactly what it will cost in the end. The costs of materials, labor, and the cost of delays are all prices that fluctuate.

Many people start building with their own cash savings, figuring that they will get a construction loan when the reserves have run out. This does not work for several reasons:

1. If you are using credit cards to fund the project, your credit score may drop, making it difficult to get further funding.

2. Bankers require a certain amount of cash reserves before they will make a loan. If you put all of yours into the project, they will not see viable cash reserves.

3. The finishing work takes the bulk of the cost, so owners are lured into a false sense of security once the home is under roof. Then, when the cash runs low, they are unable to gather more funds, which makes the home unable to pass inspection.

To offset these problems, make sure you have financing firmly in place before you break ground. That will avoid all the risk associated with loan approval. Also, be sure to borrow plenty. You can always pay it back, but it is nearly impossible to protect your losses on an underfunded project.

Bank Not Funding Draws

The bank sometimes does not draw funds when you have not completed the amount of building they feel is necessary to meet that percentage of the money. In other words, you have asked for 40 percent, but the bank only feels that you have completed 30 percent of the construction. This is difficult because you probably have subcontractors waiting to be paid.

If this happens to you, first make sure that any sections are completed that have been started. Then meet with the bank, taking along your general contractor if you have one. Make sure the bank understands what stage the project is in. Take along all paperwork, including lien releases (signed by the subs). If you cannot convince the bank to supply the extra 10 percent, consider using your own cash or credit until the percentage of completion is clearly where it needs to be for the bank to release the funds.

Project is Taking too Long

Building a house means that there will be problems. Even with the best planning and forethought, things happen. The question is, how will you handle them?

Scheduling, although emphasized so strongly in this book, is really nothing more than an estimate. The weather takes over, materials are delayed, laborers do not show, and your project is suddenly weeks behind. Rescheduling is difficult at best; a delay in one area causes a domino effect, especially if subs are not available for other time slots. Choosing to use other subs rather than wait on the ones you had picked can be a risk because the workmanship may suffer.

If the construction loan is about to expire and the project is not complete, this is another story altogether. Most construction loans will have a penalty if you go over the time limit. These can be costly — but they also can be negotiated.

As soon as you realize there is a scheduling problem, talk with the lender. Let him know exactly how much work is still needed to roll the construction loan into a permanent loan. Sometimes you can rearrange your work schedule, putting off jobs that are not necessary prior to the rollover. Then if you need an extension, go ahead and negotiate just one (a single extension will cost less than multiple extensions). If you are already past the deadline, the lender may be more likely to charge the full penalty.

Subcontractors Did Not Finish the Job

This problem occurs more frequently than you might think. Spend a quiet afternoon in your lender's office, and he will tell you of projects that drag on for years because of people starting and not finishing. This happens for a variety of reasons: subs can be unreliable, addicted to drugs or alcohol, or simply shady. Many are con artists who do this repeatedly as a way to make their living — that is why it is vital that you check references. Sometimes you and the subs do not get along; even two good personalities can become incompatible when working together. Losing a sub can cost a great deal of time and money, and the sub could even file a lien against the property, tying you up for ages.

If the subcontractor walks out, most experts say the best thing to do is let it go. Some people work hard to get the sub back on track — one woman paid to send her carpenter's entire family to Disney World after he balked at refinishing (again) her hardwood floors in a different color. However, even after the trip he was sulky and the project took twice as long as it should have.

Another owner-builder paid his framers a full payment each week. At the end of two weeks, he had paid for two full weeks of work, even though the job amounted to a little over a week's worth. The framers then did not show up for three weeks — time he spent frantically trying to get in touch with them. This knocked the entire project off schedule. When he finally spoke to the lead framer, he heard "fire me, then." The sub had no reason to return to the job. The owner had to find a new sub.

If this happens to you, try to find a contractor who is willing to step in and finish the job. Take bids and examine his background just as you did the first contractor. Consider returning to the contractors you talked with before, to see if they have time in their schedule. You will most likely have to pay a little more than you would like at this point, just to go forward and get the job done. There may be compromises that you have to make — like price and scheduling — to get the house back on track at all.

Money Ran Out

Completely running out of money is a dangerous place to be when building a home — especially if you have already sold your existing dwelling. There are many reasons why people run out of funds. Be sure to read the final list in this chapter to help save costs. But if you run out of money here are a few quick fixes to try:

Use credit cards — this is money you can pay back easily once construction is complete.

Borrow from a family member — ask for a short-term loan, and be willing to sign a promissory note. Do not sign over part ownership of the property; just promise to pay back the money on time.

Take out a home equity loan against another property — the house you live in, for example, or a vacation home. If you have not tapped into other properties, now is the time.

Borrow against other large assets — any asset you own, even stocks or jewelry, can produce a viable source of funding for the project. Do not be afraid to put up assets for short-term solutions. On the other hand, do not use this type of borrowing for long-term cost overruns.

Borrow against retirement accounts — This does not mean liquidate them, although you could as a last resort. Instead, use a credit line that is secured against your account. Liquidating retirement accounts has other ramifications, like penalties and taxes, so check with an accountant before using this method to move the project forward.

This Project is Difficult on my Marriage/Family

Any time-consuming project, like building a home, will add stress to the marriage and family life. The old adage "if you can last through building a house together, you can last through anything" is especially true if you are a DIY contractor.

Be sure to take this into consideration before you begin. Discuss the time factor with your spouse — someone will have to devote 10 to 20 hours per week to the project for the duration. This could be you, your spouse, or both. Be clear about who will carry out which aspects of house building and home chores for the duration.

During the Project

⚒ Find ways to make it fun to work on together

⚒ Allow each other time to gripe and decompress, every day

⚒ Understand the way your partner deals with stress and honor it

⚒ Take a time-out from building or renovation and focus on each other if necessary — even taking a vacation in a local hotel may seem like a luxury during a remodel!

THE LESSER EVILS

In addition to the major mistakes listed above, there are minor errors that can end up costing an owner-builder a great deal of money. Before deciding to build a home, be sure that you have addressed each of the following:

The Lot

Before purchasing the lot, did you verify that it really is worth the price? Do not take a realtor or seller's word for the lot's value; get comparable prices and determine your own.

Many subdivisions are so full of restrictions that new owners find they cannot even live in the neighborhood, much less build a dream home there. Do not allow the HOA or review committee to tell you their covenants are "standard." There is no such thing as a standard set of covenants. Many neighborhoods have restrictions on building extra garages, landscaping, and even where you park your car at night. You also will not be allowed a broad interpretation of their covenant language, so be sure you let the committee interpret any restrictions that concern you.

Utilities

Utility hookups should be free, but if you extend beyond a certain distance (perhaps 150 feet) the charge can be astronomical. This can translate into thousands of dollars — so if you find a lot that is $25,000 less than the surrounding ones, be sure to ask yourself why.

Percolation (Perc) Tests

Some buyers do not understand the way perc tests work. Perc tests must be done according to the house they are planning to build and the exact location of the future septic system. That means a generic perc test has virtually no value to a buyer. If the seller tells you a perc test has been done, ask to see the Health Department certificate. Make passing the test — your own test — a contingency of the sale.

Overgrown Vegetation

When a lot is covered with dense growth, it is nearly impossible to evaluate the soil condition, view the rock formations, study the slope, or determine how it will drain. The costs to clear a densely grown property can be extreme, especially if you are required to haul the waste to a landfill. Once cleared, the property might not be as suitable as once believed.

Spending too Much Time Shopping

You can over-shop for many items. The first is searching the Internet in a frenzy for the perfect set of plans, followed by the least expensive AND perfect set of plans. There are millions of plans available, and using the Internet is a good place to get ideas. But the cost of an original set of plans created by a reputable designer might be very close to that of the Internet copy. Besides, many Internet plans do not provide all the details, so the subs are not able to use specific drawings for their work.

The next incessant shopping spree takes place in the bidding process. If you spend too much time digging around and looking for the lowest bid, you may also obtain the contractors who have trouble getting work due to poor work habits or sub-quality work. They may also cut costs on their own expenses — insurance, licenses, illegal aliens — which can spell added costs to you in the end.

To shop less for the low bidder, know the amount of work hours the job should take, and learn what a fair market price might be. Then get three to six good bids — and make a choice.

Another shopping point is with materials. Although the ideal situation would be to find everything on sale someplace, unless you are able to spend many days driving and searching you will not find all the lowest costs on materials. Instead of chasing down sales that are located a six-hour drive away, work at building relationships with your own vendors or suppliers who can extend you discounts and negotiate prices for you.

GETTING IT WRONG

As an inexperienced builder, you are bound to make some mistakes. It is best to avoid them by knowing as much as you can at each step before taking it. For example, when you enter the local DIY store to buy supplies, you should already know exactly what materials you need. Do not rely on the (also inexperienced) store employees to recommend, say, mortar. Perhaps they'll mention the wrong one for your particular job. Their suggestion may require more drying time than you wanted or allow cracks to form. Be sure that you know what you need — and that you get it.

Problem solving during construction is an opportunity for subs to change orders or add to your contract. To offset being taken advantage of, do not rely too much on the subcontractors to create options. Each time a problem crops up, research it with the intent to find your own solution — then ask the sub for his input. That way you will know whether the solution he offers is viable.

CONTROLS

Owner-builders notoriously underestimate time and costs. Often up to 10 percent of the job is missing from the master list. This renders the builder incapable of creating an efficient, cohesive plan. There is an optimal sequence of work for every job and it varies from one project to another. If things are done out of sequence, there may be additional work to pay for. Take the time to schedule very efficiently, and mentally run through the plan before making it a reality.

Conclusion: A Final Word on Saving Money

To save money, be alert at all times to cost-cutting solutions. There are some instances in which you do not want to scrimp. These instances are:

- Buy the best lot you can afford.
- Hire the best contractors, especially for the foundation, framing, and carpentry.
- Do not sign a contract until you are sure that every detail is included, and you are satisfied with all the terms.

One of the best ways that you can save is to do work yourself. Offer to work as a helper on certain steps of the construction. Subcontractors who do not have a regular helper will allow you to do this, and you will get on-the-job experience as well as save on costs. Consider doing some jobs like:

- Painting
- Wallpapering

- Installing doors and light fixtures
- Trim work
- Cleanup

When determining costs during pre-construction, always choose a bid over an estimate as a starting point. Using estimates can throw the budget way off, especially if you have forgotten something.

Consider downgrading materials that will not be seen; use prefabricated items if they are available, windows with fake pane dividers, and even framing.

Use a rectangular floor plan, carefully locating partitions to intersect the exterior wall studs to save on materials.

Cluster rooms that require plumbing fixtures, like bathrooms, the kitchen, and utility rooms. For a two-story home, place the upper level plumbing directly over the lower level plumbing.

Consider using engineered floor trusses instead of standard lumber; eliminate double floor joists and band joists as well as bridging (which is no longer required for joists less than 2x12).

Select high-efficiency HVAC units with a high EER, and Energy Star® appliances. These will save money in just a few months' bills.

In going over the final punch list, there may still be many small items that must be completed. There could be dings in the wall, paint that needs touching up, windows that need to be lined up, and so on. If your contract calls for the subcontractor to come back and adjust or repair the work, be sure to get them back out soon. If not, these are items that you can take care of yourself.

After completion, property tax will go up almost right away. Homes are assessed according to value; normally you have about 60 days to challenge the assessment. Do not be unfriendly to the assessor, but be aware that you can refuse to allow them inside the home in many areas. When the assessment arrives, consider challenging the bill if comparable homes are appraising for less or if your actual costs were less. Back up your claim with your receipts.

The information presented in this book is meant to help you get started in the process, and help you have the confidence to build your own home. Good luck to you — and please feel free to write to me about your experience in building or remodeling your dream home!

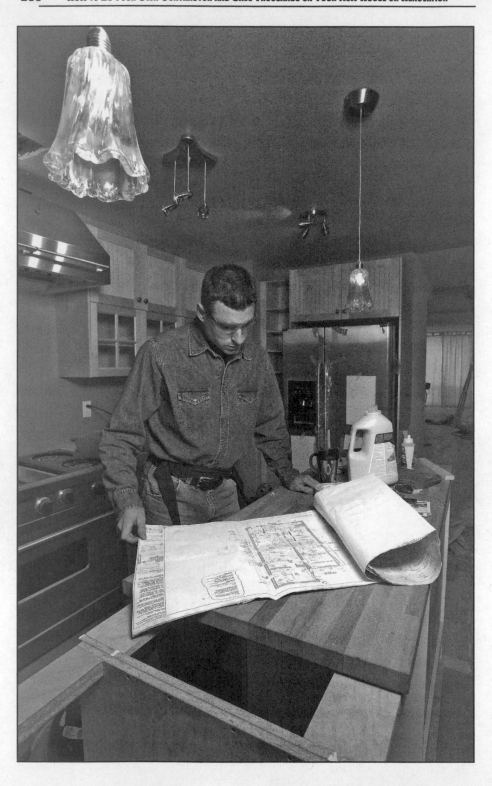

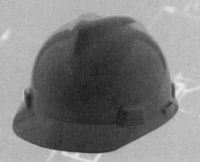

Appendix: Resources

WEB SITES

Searching for Land

Lotfinders — **www.lotfinders.com**

This is an Internet site dedicated to helping consumers find land for building a custom home. They actively work with developers and other landowners so that they have plenty of listings. They can also help you find out who owns a particular property if you are interested in making an offer but cannot find the owner.

Land Auction Bid   **www.landauctionbid.com**

This is a live auction which is Web cast, so that you can register and bid online. The largest land auction company in the country, they offer building lots as well as commercial land for sale all over the country. The company also offers financing on most of the lots.

House Plans and Design

Hanley Wood — **www.eplans.com**

The Garlinghouse Company — **www.garlinghouse.com**

Good Housekeeping Home Plans Magazine

Building Green — **www.BuildingGreen.com**

Financing

Indy Mac — **http://www.Indymacbank.com**

Construction Loan Center — **http://www.constructionloancenter.com/**

Custom Mortgage Corporation — **http://www.custommtgcorp.com/**

Building

Do It Yourself — **http://www.doityourself.com**

An Internet resource created to give practical advice on every sort of DIY project. It also offers links to suppliers and an online discussion forum.

Residential Design & Build Magazine — **http://www.rdbmagazine.com/**

This is an online magazine that is geared toward professional builders and architects, but it is full of good ideas for owner-builders, too.

Owner Builder Network — **http://www.ownerbuildernetwork.com/**

This site was created specifically for owner-builders, primarily Texas residents.

This Old House — **http://www.thisoldhouse.com**

Like its sister magazine, This Old House offers articles and a videos about every imaginable remodeling project.

PERIODICALS

Builder

Custom Home

Fine Home-Building

Home

Home and Architectural Trends

Homestyles Home Plans

Southern Accents

Taunton's Fine Home Building

This Old House

ASSOCIATIONS

The following associations are great resources for information on their members as well as for locating professionals in the field:

American Institute of Architects (AIA) — **http://www. aia.org**

National Association of Home Builders (NAHB) — **www.nahb.org**

OTHER GOOD STUFF

These are materials that you probably will want to have on hand before beginning construction:

Your local Building Code — This is obtainable from the local building inspector. It will outline all the rules that must be adhered to when constructing or remodeling a home.

HUD Guidelines to minimum Property Standards for Single Family Housing — This is especially helpful in areas of the country that have not adopted a Building Code. HUD requires that all properties insured with an FHA mortgage meet a code, so if you ever plan to sell the property it is good to have a copy of these codes. Obtain it from:

Department of Housing and Urban Development
FHA Standards, office of Manufactured Housing Programs
451 7th St. SW Room 9168
Washington, DC 20410-8000
Phone 202-708.6423

Author Acknowledgements & Biography

I could never have written this book without the help and guidance of so many professionals in the field. Homeowners, builders and investors have patiently responded to my numerous questions, e-mails, and late-night calls without complaint. No one has ever refused to answer or wondered why I needed to know at 11 o'clock at night.

As always, I thank Atlantic Publishing, and particularly Angela Adams, senior editor. No book is successful without a keen editor and Angela is one of the best.

I thank my family, who often has to put up with eating crackers and cheese (again) because Mom's working on her book. They're always willing to listen to a paragraph or two "just to see if it sounds funny." My mom says she is going to start her own construction company soon, after reading all these books.

Most of all, I want to give a special thanks to my partner, M., who has given me support, editing help, and wrote most of the checklists you'll find in this book. His contracting savvy and enthusiasm kept me going at that critical point when the book's nearly done, but you'd like to put it down for a while.

Readers of my earlier books played an essential role in how this book was laid out. You have requested more stories and anecdotes, and that is what I have tried to present in *Be Your Own Contractor*.

Tanya R. Davis is the author of *The Real Estate Developer's Handbook*, *How to Start a Financially Successful Construction Company*, and co-author of *How to Make a Fortune at Real Estate Auctions*. She is also a weekly news columnist. Visit her Web site at **http://www.tanyardavis.com.**

House Building Glossary Terms

Abstract of Title — A summary of recorded transactions concerning a property.

Acrylic Resin — A thermoplastic resin used in latex coatings.

Accrued Interest — Interest which has been incurred but is not yet paid.

Adjustable Rate Mortgage (ARM) A mortgage in which the interest rate is adjusted periodically based on a pre-selected index.

Adjustment Interval or Adjustment Period — The length of time between rate adjustments on an Adjustable Rate Mortgage (ARM).

Aggregate — Irregularly shaped gravel suspended in cement.

Air Chamber — A pipe appendage with trapped air that is added to a line. Serves as a shock absorber to slow or eliminate air hammer.

Air-dried lumber — Dried via exposure to air without artificial heat.

Alkyd Resin — One of a large group of synthetic resins used in latex paint.

Amperage — The amount of current flowing in a wire.

Amortization — The gradual reduction of a loan debt through periodic installment payments of principal and interest.

Anchor — Irons that hold together timbers or masonry.

Anchor Bolt — bolt that secures a wooden sill plate to concrete or masonry floor or foundation wall.

Apron — Trim that is used at the base of windows; also used as the base to build out crown molding.

Assumability — A loan feature that allows the loan to be transferred from the seller to the purchaser of a new home with the same terms and conditions, subject to lender approval.

Attic Ventilators — In houses, screened openings that ventilate an attic space.

Backfill — Process of placing soil against a foundation after all necessary foundation processes have been performed.

Backflow — The flow of water or other fluids into the main source of potable water.

Balloon Frame — The lightest and most economical form of construction. Studs and corner posts are continuous lengths from the first floor line or sill to the roof plate. The upper story floor joists are carried on ledgers or girts let into the studs.

Balloon Mortgage — A short-term, fixed-rate loan with low payments for a set number of years and a large balloon payment of the remainder of the principal due at the end of the term.

Baluster — A vertical support in a railing system; may be used on stairs, balconies, and porches.

Balustrade — A railing system made of balusters, top rails, and possibly a bottom rail.

Base Shoe — Molding applied next to the floor on the interior baseboards, usually a quarter round strip.

Batten — Narrow strips of wood covering joints, such as on plywood or wide boards.

Batter Board — A pair of horizontally place boards nailed to vertical posts.

Batter Pile — Pile driven at an angle to brace a structure.

Bazooka — An automated drywall application tool. It tapes, muds, and feathers in one pass.

Bead — A corner or edge that must be finished with stucco.

Bed Molding — The trim piece that covers the intersection of the vertical wall and any horizontal overhanging surface, like a soffit. In a series of moldings, the bed molding is the lowest one.

Berm — A mound of built-up dirt used for either drainage or landscaping.

Bevel Board — Also called a pitch board, used in framing a roof or stairway to lay out bevels.

Bibb — Special covering used where lines run through an exterior wall.

Bi-weekly Mortgage — A payment plan under which the borrower pays one half of a monthly payment every two weeks.

Blanket Mortgage — A mortgage covering at least two or more pieces of real estate.

Blind Nailing — Nailing so that the nail heads are not visible on the face of the work. Used at the tongue of matched boards.

Board Foot — Unit of measure equal to a 1" thick piece of wood, 1' square. Length x Width x Thickness equals board feet.

Boarding In — The process of nailing boards on the outside studs of a house.

Bolster — A short horizontal beam on top of a column. It supports and decreases the girder span. May be wood or steel.

Bond — A document representing a right to certain payments on underlying collateral.

Boston Ridge — A method of putting asphalt or wood shingles at the ridge or hip of a roof for a finished look.

Break Joints — A system of arranging joints so that they are not directly over adjoining joints. Done in shingles, siding, brickwork.

Breather Paper — Paper that lets water vapor pass through.

Bridge Loans — Financing for homeowners who are constructing a new owner-occupied primary residence but are keeping the current owner-occupied residence, which they intend to sell.

Broker — An independent middleman who arranges deals between borrowers and lenders.

Built-up Roof — A roof that is composed of 3 to 5 layers of asphalt felt, laminated with coal tar or asphalt. The top is crushed slag or gravel.

Buy-Down — A situation in which the seller of a property contributes money, allowing the lender to give the buyer a lower rate and payment, typically in exchange for an increase in sales price.

Cant Strip — A piece of lumber used at the junction of a deck, or roof, and a wall to modify the angle.

Cantilever — A horizontal structure component that projects from a building, such as a step, balcony, beam or canopy, that is without external bracing and appears to be self-supporting.

Cap (Construction) — Hardware that terminates plumbing lines; or the upper member of a column, door cornice, or molding.

Casement Window — A window that swings out to the side on hinges.

Cash Out — A refinance for more than the balance of the current mortgage to take cash out from the loan at closing.

Casing Nails — Used to apply finish trim and millwork; the nail head is set below the surface of the wood so that it is hidden.

Chamfer — The beveled edge of a board.

Clean-out — A sealed opening in a pipe that can be removed to clean out a clog.

Clear Title — Real property ownership free of liens, defects, encumbrances, or claims

Cleat — A length of wood affixed to a surface to give a firm foothold, or to hold an object in place.

Closing (or Settlement) — Meeting between the buyer, seller, and lender or their agents, at which property ownership and funds legally change hands.

Cloud on Title — An outstanding claim or

encumbrance that, if valid, would affect or impair the owner's title.

Combined Loan To Value — The percentage of the property value borrowed through a combination of more than one loan.

Collar Beam — Boards 1" or 2" thick connecting roof rafters near the ridge board in order to strengthen the roof structure.

Convertible ARMs — ARMs with the option of conversion to a fixed loan during a given time period.

Corbel — Extending a course of bricks beyond the face of a wall. The total projection should never exceed the thickness of the wall.

Cornice — Trimwork finishing the intersection of the roof with the siding; consisting of a fascia board, a soffit, and appropriate moldings.

Cornice Return — The underside of the cornice; trim, transitioning from the horizontal eave line to the sloped roofline.

Cost of Funds Index (COFI) — A common index used in adjustable rate loans based on the weighted-average interest rate paid for deposits by savings institutions that are members of the 11th Federal Home Loan Bank District.

Course of Construction — Insurance policy in the form of an "all risk" policy with fire, extended coverage, builder's risk, replacement cost, vandalism, and malicious mischief insurance coverage.

Covenants, Conditions, and Restrictions (CC&Rs) — A document that defines the use, requirements, and restrictions of a residential area, condominium, or Planned Unit Development (PUD).

Countersink — To set the head of a nail below the surface.

Cove Molding — Has a concave face; used as trim or for finishing interior corners.

Crimp — A crease formed in sheet metal to make it less flexible or to make it easy to fasten onto.

Cripple — A stud used for bracing under structural framing, like windows.

Crown Molding — Trim piece applied at the intersection of ceiling and walls in formal rooms.

Crusher Run — Crushed stone with sharp edges, used for driveway and foundation support as a base. Packs down to give a stable surface.

Dado — A rectangular groove across the width of a board or plank.

Deferred Interest — Amount added to the balance of a loan when monthly payments are insufficient to cover the interest incurred.

Door Jamb — The surrounding case into which a door closes. It consists of two upright side jambs and a horizontal head jamb.

Dormer — An opening in a sloping roof that projects out to form a wall suitable for windows or other openings.

Dovetail — Joint made by cutting pins in

the shape of dove tails which fit between matching cuts on another piece of wood.

Draw — During construction, a request for a percentage of the funds by the builder.

Earnest Money — Deposit made by a buyer toward the down payment as evidence of good faith when the purchase agreement is signed.

EER (Energy Efficiency Rating) — A national rating system that must be displayed on appliances that measures the efficient use of electrical power.

Eaves — The part of the roof that extends beyond he outside walls of a house.

Effective Interest Rate — The cost of a mortgage expressed as a yearly rate, usually higher than the interest rate on the mortgage since this figure factors into the upfront costs of acquiring the loan.

Equity — The difference between the current market value of a property and the outstanding mortgage balance.

Escrow — An amount set up by the lender into which the borrower makes periodic payments, such as monthly, for taxes, hazard insurance, assessments, and mortgage insurance premiums.

Expansion Joint — A bituminous fiber strip that separates blocks or concrete to allow for expansion (due to temperature changes) without cracking.

Face Nailing or Direct Nailing — Nailing perpendicular to the board or other material.

Fascia — A flat board covering the end of the rafter, or the board that connects the top of siding to the bottom of the soffit.

Feathering — Applying drywall compound in successive coats, thereby widening the compound joint.

Ferrule — Aluminum sleeve used to attach trough to gutter spike.

Fill Dirt — Loose dirt brought in from another location to use under slabs, driveways, and sidewalks; sturdier than topsoil.

Finish Grade — The process of leveling and smoothing topsoil into its final position.

Fire Stop — A solid, tight closure of a concealed space in order to prevent fire from spreading. In a frame wall, this usually requires 2x4 cross blocking between studs.

Fishplate — A wood or plywood piece used at the junction of opposite rafters near the ridge line.

Fixed-Rate Mortgage — A mortgage in which the interest rate does not change for the life of the loan. Payments are also fixed.

Flashing — Galvanized sheet metal used for a lining around joints, normally between shingles and chimneys, exhaust or ventilation vents, and other protrusions that might allow water seepage under the shingles.

Flue — The passage in the chimney through which smoke, gas, and fumes ascend.

Furring — Long strips of wood attached to walls or ceilings to allow attachment of drywall or ceiling tiles. "Furring out" refers to using the furring strips to bring the wall further into the room.

GFCI (Ground Fault Circuit Interrupter) — An extra-sensitive circuit breaker used for additional protection against shock.

Gambrel — A roof that slopes steeply at the edge of a building, changing to a shallower slope at the center.

Gem Box — Metal box installed during the electrical rough-in that holds outlets and receptacles.

Graduated Payment Mortgage (GPM) — Mortgage in which initial low payments (with potential negative amortization) increase regularly for several years and then level off.

Gusset — A flat wood, plywood, or similar piece providing a connection at the intersections of wood.

Gypsum Plaster — A mix of gypsum that is made to be sued with sand and water added for base coating plaster.

H Clip — A metal clip for holding adjacent plywood sheets in alignment.

Hazard Insurance — A policy that protects the insured against loss due to fire or certain natural disasters in exchange for a premium paid to the insurer.

Head Lap — Vertical length in inches of the amount of overlap between shingles.

Heel of a Rafter — End or foot that rests on the wall plate.

Heel Wedges — Triangular shaped wood that is driven into gaps between rough framing, like window framing, to give solid backing.

Hip Roof — A roof sloping upward toward the center from every side; requires a hip rafter at each corner.

Home Equity Loan — An additional mortgage secured by the equity in the home.

Hopper Window — Window that is hinged at the bottom and swings inward.

Housing Code — Local government ordinance that sets minimum standards of safety and sanitation for existing residential buildings.

I-beam — Steel beam with a cross section so that it is shaped like a capital I; used for long spans over wide openings where there are heavy roof loads imposed above the opening.

Initial Rate — The rate charged during the first interval of an adjustable rate mortgage.

Insulation — Any material used to reduce the effects of heat, cold, or sound transmission and to reduce fire hazard.

Insulative Panel — Flat sheet material made of insulative material to improve thermal resistance characteristics of assembly.

Insolvency — Condition of a person unable to pay debts as they fall due.

Interest — Charge paid for borrowing money.

Investor Rehab Financing—This program is designed to provide a loan for investors to acquire and rehabilitate a property for future rental use.

Involute — Curved part of trim that terminates a piece of staircase railing in traditional homes.

Isolation Joint — A joint in which two incompatible materials are isolated from each other to prevent chemical action between the two.

Jack Rafter — A shortened rafter that joins a hip or valley to the top of a wall plate.

Jack Post — A hollow metal post with a jack screw in one end to adjust the height.

Jack Stud — A short stud that does not extend from floor to ceiling; may reach from floor to a window, for example.

Jalousie Window — A window with stationary or adjustable blinds angled to permit air and provide shade, at the same time preventing rain from entering.

Jamb — The vertical side of a door or window. The side piece or post of an opening; commonly applied to the door frame.

Joint Clip — Fastener used vertically, sharp edges down, over edges of two pieces, then hammered down into them.

Joist — One of a group of light, closely spaced beams used to support a floor deck or flat roof.

Joist Hanger — A metal stirrup that supports the ends of joists so that they are flush with the girder.

Joule — In the international system of units, the amount of energy needed to raise one kilogram of water one degree Celsius.

Journeyman — An experienced, reliable worker who has learned his trade and works for another person.

Junction Box — A metal box in which runs of cable meet and are protectively enclosed.

Kerf — The area of a board removed by the saw; vertical notch or cut in a batter board where a string is fastened.

K Bracing — The form of bracing where a pair of braces located on one side of a column terminates at a single point within the clear column height.

Keyways — Tongue and groove connection where perpendicular concrete components meet; designed to prevent movement between components.

Keyed Joint — A joint in which one structural member is keyed or notched into an adjoining member as in timber construction. In masonry construction, a finished joint of mortar which has been tooled concave.

Kick Plate — A metal strip or plate that runs along the bottom edge of a door to protect against the marring of the finished surface.

Kiln Dried — Lumber dried by means of controlled heat and humidity.

Lag Screws — Large screws with heads designed to be turned with a wrench.

Laminate — Thin material, often plastic or wood, glued to the exterior of a cabinet.

Landing — A platform between flights of stairs.

Lath — A grid applied to exterior sheathing used as a base for stucco.

Lender's Contingency — This is a reserve to cover unforeseen circumstances in the construction of the new home.

Lien — A legal claim against a property that must be paid when the property is sold.

Lifetime Interest Rate Cap — The highest interest rate that can be charged for an adjustable rate mortgage during the life of the loan.

Lintel — The horizontal beam placed over an opening.

Loan Origination Fee (or Processing Fee) — Fee charged by a lender that compensates for the work in evaluating and processing the loan.

Lock (or Lock In) — A lender's guarantee of an interest rate and related points for a set period of time, usually between loan application and loan closing.

Loan To Value (LTV) Ratio — The percentage of the property value borrowed (loan amount/property value, loan to value ratio).

Luminaire — A complete lighting unit consisting of a light source, switch, globe, reflector, housing, and wiring.

Mansard Roof — A type of roof that slopes very steeply around the perimeter of the building to full wall height.

Margin — The percentage amount added to an index to calculate the interest rate of an adjustable rate mortgage at each adjustment.

Mastic — Pasty material used for setting tile or for protective coating for thermal insulation or waterproofing.

Millwork — Generally, all materials made of finished wood (doors, windows, door frames, blinds, moldings) and manufactured in millwork plants are called millwork.

Moisture Barrier — A membrane used to prevent the migration of liquid water through a floor or wall.

Moisture Movement — The movement of moisture through a porous medium; the effects of such movement on efflorescence and volume change in hardened cement paste, mortar, concrete, or rock.

Mono Pitch Truss — A truss that would develop a shed type roof.

Mortgage — Document creating a lien on a property as security for the payment of a debt.

Mortgage Banker — A lender that originates and funds, then sells and services mortgage loans.

Mortgage Broker — A person or entity that

arranges financing for borrowers, but places loans with lenders rather than funding them with the broker's own money.

Mortar — A mixture of cement paste and fine aggregate; in fresh concrete.

Mortise — A slot cut edgewise into a board to receive a tenon of another board to form a joint.

Mullion — A vertical divider in the frame between windows, doors, or other openings.

Muntin — A short bar separating glass panes in a window sash.

Nail Inspection — An inspection made by a municipal building inspector after the drywall material is hung with nails and screws (and before taping).

NEC (National Electrical Code) A set of rules governing safe wiring methods. Local codes — which are backed by law — may differ from the NEC in some ways.

Neutral Wire — Color-coded white, this carries electricity from an outlet back to the service panel.

Newel Post — The large starting post to which the end of a stair guard railing or balustrade is fastened.

Nonbearing Wall — A wall supporting no load other than its own weight.

Non-Assumption Clause — A statement in a mortgage contract forbidding the assumption of the mortgage by another borrower without the prior approval of the lender.

Non-dischargeable Debt — Debt, such as taxes, that cannot be forgiven in a bankruptcy liquidation.

Notice of Default — Written notice to a borrower that a default has occurred and that legal action may be taken.

Nosing — The projecting edge of a molding or drip or the front edge of a stair tread.

Notch — A crosswise groove at the end of a board.

Nozzle — The part of a heating system that sprays the fuel of fuel-air mixture into the combustion chamber.

O.C. (On Center) — The measurement of spacing for studs, rafters, and joists in a building from the center of one member to the center of the next.

Oakum — Loose hemp or jute fiber that is impregnated with tar or pitch and used to caulk large seams or for packing plumbing pipe joints

Open Hole Inspection — When an engineer (or municipal inspector) inspects the open excavation and examines the earth to determine the type of foundation (caisson, footer, wall on ground) that should be installed in the hole.

Oriented Strand Board or OSB A manufactured 4'x8' wood panel made out of 1" to 2" wood chips and glue. Often used as a substitute for plywood.

Outrigger — An extension of a rafter beyond the wall line. T

Pad Out, Pack Out — To shim out or add strips of wood to a wall or ceiling in order that the finished ceiling/wall will appear correct.

Panel — A thin flat piece of wood, plywood, or similar material, framed by stiles and rails.

Paper, Sheathing — Paper or felt used in wall and roof construction to protect against air and moisture passage.

Parapet — A wall placed at the edge of a roof to prevent people from falling off.

Parting Stop or Strip — A small wood piece used in the side and head jambs of double hung windows to separate the upper sash from the lower sash.

Paver, Paving — Materials, commonly masonry, laid down to make a firm, even surface.

Pedestal — A metal box installed at various locations along utility easements that contain electrical, telephone, or cable television switches and connections.

Penalty Clause — A provision in a contract that provides for a reduction in the amount otherwise payable under a contract to a contractor as a penalty for failure to meet deadlines or for failure of the project to meet contract specifications.

Per Diem Interest — Interest calculated per day. Depending on the day of the month on which closing takes place, borrower pays interest from the date of closing to the end of the month.

Percolation Test (Perc Test) — Tests that a soil engineer performs on earth to determine the feasibility of installing a leach field type sewer system on a lot.

Perimeter Drain — 3" or 4" perforated plastic pipe that goes around the perimeter (either inside or outside) of a foundation wall (before backfill) and collects and diverts ground water away from the foundation.

Pigtails, Electrical — The electric cord that the electrician provides and installs on an appliance such as a garbage disposal, dishwasher, or range hood.

Pilot Hole — A small-diameter, pre-drilled hole that guides a nail or screw.

Pilot Light — A small, continuous flame (in a hot water heater, boiler, or furnace) that ignites gas or oil burners when needed.

Pitch — The incline slope of a roof or the ratio of the total rise to the total width of a house.

PITI — Principal, interest, taxes, and insurance (the four major components of monthly housing payments).

Plan View — Drawing of a structure with the view from overhead, looking down.

Plate — Normally a 2x4 or 2x6 that lays horizontally within a framed structure, such as: **Sill Plate** — A horizontal member anchored to a concrete or masonry wall; **Sole Plate** — Bottom horizontal member of a frame wall; **Top Plate** — Top horizontal member of a frame wall supporting ceiling joists, rafters, or other members.

Plenum — The main hot-air supply duct leading from a furnace.

Plot Plan — An overhead view plan that shows the location of the home on the lot.

Plough, Plow — To cut a lengthwise groove in a board or plank.

Plumb — Exactly vertical and perpendicular.

Plumb Bob — A lead weight attached to a string. It is the tool used in determining plumb.

Plumbing Boots — Metal saddles used to strengthen a bearing wall/vertical stud(s) where a plumbing drain line has been cut through and installed.

Plumbing Ground — The plumbing drain and waste lines that are installed beneath a basement floor.

Plumbing Jacks — Sleeves that fit around drain and waste vent pipes at, and are nailed to, the roof sheeting.

Plumbing Rough — Work performed by the plumbing contractor after the Rough Heat is installed.

Plumbing Stack — A plumbing vent pipe that penetrates the roof.

Plumbing Trim — Work performed by the plumbing contractor to get the home ready for a final plumbing inspection.

Plumbing Waste Line — Plastic pipe used to collect and drain sewage waste.

Plywood — A panel (normally 4'x8') of wood made of three or more layers of veneer, compressed and joined with glue, and most often laid with the grain of adjoining plies at right angles to give the sheet strength.

Point Load — A point where a bearing/ structural weight is concentrated and transferred to the foundation.

Points (or Discount Points) — Money paid to a lender at closing in exchange for a lower interest rate.

Portland Cement — Cement made by heating clay and crushed limestone into a brick and then grinding to a pulverized powder state.

Post — A vertical framing member typically designed to carry a beam.

Post-and-Beam — A basic building method that uses just a few hefty posts and beams to support an entire structure.

Power of Attorney — Legal document authorizing one person to act on behalf of another.

Prepaid Interest — Interest charged to a borrower at closing to cover interest on the loan between closing and the end of the month in which the loan closes.

Prepayment — Full or partial payment of the principal before the due date.

Pressure Relief Valve (PRV) — A device mounted on a hot water heater or boiler which is designed to release any high steam pressure in the tank to prevent tank explosions.

Pressure-treated Wood — Lumber that has been saturated with a preservative.

P Trap — Curved, "U" section of drain pipe that holds a water seal to prevent sewer gasses from entering the home through a fixtures water drain.

PUD (Planned Unit Development) A project or subdivision that includes common property that is owned and maintained by a homeowners' association for the benefit and use of the individual PUD unit owner.

Pump Mix — Special concrete that will be used in a concrete pump. The mix normally has smaller rock aggregate than regular mix.

Punch List — A list of discrepancies that need to be corrected by the contractor.

Punch Out — To inspect and make a discrepancy list.

Putty — A type of dough used in sealing glass in the sash, filling small holes and crevices in wood, and for similar purposes.

PVC or CPVC (Poly Vinyl Chloride) A type of white or light gray plastic pipe sometimes used for water supply lines and waste pipe.

Quarry Tile — A man-made or machine-made clay tile used to finish a floor or wall. Generally 6"x6"x1/4" thick.

Quarter Round — A small trim molding that has the cross section of a quarter circle.

Rabbet — A rectangular longitudinal groove cut in the corner edge of a board or plank.

Radiant Heating — A method of heating, which can consist of a forced hot water system with pipes placed in the floor, wall, or ceiling.

Radon — A naturally-occurring, heavier than air, radioactive gas common in many parts of the country.

Rafter — Lumber used to support the roof sheeting and roof loads.

Rafter, Hip — A rafter that forms the intersection of an external roof angle.

Rail — Cross members of panel doors or of a sash.

Rake — The angled edge of a roof, at the end where it passes the gable.

Rake Fascia — The vertical face of the sloping end of a roof eave.

Rebar — Metal rods used to improve the strength of concrete structures.

RESPA (Real Estate Settlement Procedures Act) — Law requiring lenders to give borrowers advance notice of closing costs.

Reclamation — The right of the person with title to a property to recover it from the debtor in the event of a bankruptcy.

Re-Conveyance — The transfer of property back to the owner when a mortgage is fully repaid.

Refinancing — The process of paying off one loan with the proceeds from a new loan secured by same property.

Repossession (Foreclosure) — Legal process by which the lender forces the sale of a property because the borrower has not met the mortgage terms.

Reflective Insulation — Sheet material with one or both surfaces emitting low heat, such as aluminum foil.

Resorcinol — An adhesive high in both wet and dry strength.

Retainage — A percent of payment that is held back to ensure a job is completed.

Return — Ductwork that leads back to the HVAC unit.

Reverse Board and Batten — Narrow battens nailed vertically to wall framing; wider boards are nailed over these to form siding.

Ribbon (Girt) — Normally a 1x4 board horizontally supporting ceiling or second-floor joists. Set within the studs.

Rolled Roofing — A roofing material composed of fiber.

Roof Sheathing — The boards or sheets fastened to the rafters, on which shingles or other roofing material is laid.

Rottenstone — An abrasive stone used for rubbing a transparent finish to give it a smooth surface.

Rout — The removal of material by cutting or gouging a groove.

Rubble Masonry — Uncut stones used for foundations and other rough work.

Run — In stairs, the front to back width of a step or a flight of stairs.

R Value — A measure of insulation. A measure of a materials resistance to the passage of heat.

Saddle — A small second roof built behind the back side of a fireplace chimney or other part of the roof to divert water around the chimney.

Sack Mix — The amount of Portland cement in a cubic yard of concrete mix.

Sand Float Finish — Lime that is mixed with sand, resulting in a textured finish on a wall.

Sanitary Sewer — A sewer system designed for the collection of waste water from the bathroom, kitchen, and laundry drains, and is not necessarily designed to handle storm water.

Sash — A single light frame containing one or more lights of glass.

Sash Balance — A device, usually operated by a spring and designed to hold a single hung window vent up and in place.

Scab — Short length of board nailed over the joint of two boards that butt end to end.

Scantling — Lumber with a cross section from 2x4" to 4x4".

Scarfing — A joint between two pieces of wood that allows them to be spliced lengthwise.

Scotia — Hollow molding used for part of a cornice.

Scrap Out — The removal of all drywall material and debris after the home is installed with drywall.

Screed, Plaster — A small strip of wood, the thickness of the plaster coat, used as a guide for plastering.

Scribing — Cutting and fitting woodwork to an irregular surface.

Scupper — (1) An opening for drainage in a wall, curb or parapet, (2) The drain in a downspout or flat roof, connected to the downspout.

Second Mortgage — A subordinate mortgage made in addition to a first mortgage.

Self-sealing Shingles — Shingles containing factory-applied strips or spots of self-sealing adhesive.

Septic System — An on-site waste water treatment system.

Service Entrance Panel — Main power cabinet where electricity enters a home wiring system.

Service Equipment — Main control gear at the service entrance, such as circuit breakers, switches, and fuses.

Service Lateral — Underground power supply line.

Settlement (Closing) — Meeting between the buyer, seller, and closing agent at which property and funds legally change hands.

Settlement Sheet — The computation of costs payable at closing which determines the seller's net proceeds and the buyer's net payment.

Setback Thermostat — A thermostat with a clock which can be programmed to come on or go off at various temperatures and at different times of the day/week.

Settlement — Shifts in a structure, caused by freeze-thaw cycles underground.

Sewage Ejector — A pump used to lift waste water to a gravity sanitary sewer line.

Sewer Lateral — The portion of the sanitary sewer which connects the interior waste water lines to the main sewer lines.

Sewer Stub — The junction at the municipal sewer system where the home's sewer line is connected.

Sewer Tap — The physical connection point where the home's sewer line connects to the main municipal sewer line.

Shake — A wood roofing material, normally cedar or redwood.

Shear Block — Plywood that is face nailed to short (2x4's or 2x6's) wall studs (above a door or window, for example).

Sheathing, Sheeting — The structural wood panel covering, normally OSB or plywood, used over studs, floor joists or rafters/trusses of a structure.

ShedRoof — A roof containing only one

sloping plane.

Sheet Metal Work — All components of a house employing sheet metal, such as flashing, gutters, and downspouts.

Sheet Metal Duct Work — The heating system. Round or rectangular metal pipes and sheet metal (for Return Air) and installed for distributing warm (or cold) air from the furnace to rooms in the home.

Sheet Rock — A manufactured panel made out of gypsum plaster and encased in a thin cardboard.

Shim — A small piece of scrap lumber or shingle, often wedge shaped, which when forced behind a furring strip or framing member forces it into position.

Shingles — Roof covering of asphalt. Asbestos, wood, tile, slate, or other material cut to stock lengths, widths, and thicknesses.

Shingles, Siding — Various kinds of shingles, used over sheathing for exterior wall covering of a structure.

Short Circuit — A situation that occurs when hot and neutral wires come in contact with each other.

Side Sewer — The portion of the sanitary sewer which connects the interior waste water lines to the main sewer lines.

Siding — The finished exterior covering of the outside walls of a frame building.

Siding (Lap Siding) — Slightly wedge shaped boards used as horizontal siding in a lapped pattern over the exterior sheathing.

Sill — (1) The 2x4 or 2x6 wood plate framing member that lays flat against and bolted to the foundation wall (with anchor bolts) and upon which the floor joists are installed. (2) The member forming the lower side of an opening, as a door sill or window sill.

Sill Cock — An exterior water faucet (hose bib).

Sill Plate (Mudsill) — Bottom horizontal member of an exterior wall frame which rests on top a foundation, sometimes called mudsill.

Simple Interest — Interest computed only on the principal balance.

Single Hung Window — A window with one vertically sliding sash or window vent.

Slab, Door — A rectangular door without hinges or frame.

Slab on Grade — A type of foundation with a concrete floor which is placed directly on the soil.

Slag — Concrete cement that sometimes covers the vertical face of the foundation material.

Sleeper — A wood member embedded in concrete, as in a floor, that serves to support and to fasten the subfloor or flooring.

Sleeve(s) — Pipe installed under the concrete driveway or sidewalk, and that will be used later to run sprinkler pipe or low voltage wire.

Slope — The incline angle of a roof surface, given as a ratio of the rise (inches) to the run (in). See also pitch.

Slump — The "wetness" of concrete. A 3" slump is dryer and stiffer than a 5" slump.

Soffit — The area below the eaves and overhangs.

Soil Stack — A plumbing vent pipe that penetrates the roof.

Sole Plate — The bottom, horizontal framing member of a wall that is attached to the floor sheeting and vertical wall studs.

Solid Bridging — A solid member placed between adjacent floor joists near the center of the span to prevent joists or rafters from twisting.

Sonotube — Round, large cardboard tubes designed to hold wet concrete in place until it hardens.

Splash Block — Portable concrete (or vinyl) channel placed beneath an exterior sill cock (water faucet) or downspout in order to receive roof drainage from downspouts and to divert it away from the building.

Square-tab Shingles — Shingles on which tabs are all the same size and exposure.

Squeegie — Fine pea gravel used to grade a floor (normally before concrete is placed).

Stack (Trusses) — To position trusses on the walls in their correct location.

Standard Practices of the Trade(s) One of the more common basic and minimum construction standards.

Starter Strip — Asphalt roofing applied at the eaves that provides protection by filling in the spaces under the cutouts and joints of the first course of shingles.

Stair Carriage or Stringer — Supporting member for stair treads.

Stair Landing — A platform between flights of stairs or at the termination of a flight of stairs.

Static Vent — A vent that does not include a fan.

Step Flashing — Flashing application method used where a vertical surface meets a sloping roof plane. 6"x6" galvanized metal bent at a 90 degree angle, and installed beneath siding and over the top of shingles.

Stick Built — A house built without prefabricated parts. Also called conventional building.

Stile — An upright framing member in a panel door.

Stool — The flat molding fitted over the window sill between jambs and contacting the bottom rail of the lower sash.

Stop Box — Normally a cast iron pipe with a lid (@ 5" in diameter) that is placed vertically into the ground, situated near the water tap in the yard, and where a water cut-off valve to the home is located (underground).

Stop Order — A formal, written notification to a contractor to discontinue some or all work on a project.

Stop Valve — A device installed in a water supply line, near a fixture, that permits an individual to shut off the water supply to one fixture without interrupting service to the rest of the system.

Storm Sash or Storm Window — An extra window placed outside of an existing one, as additional protection against cold weather.

Storm Sewer — A sewer system designed to collect storm water and is separated from the waste water system.

Strike — The plate on a door frame that engages a latch or dead bolt.

Strip Flooring — Wood flooring consisting of narrow, matched strips.

Structural Floor — A framed lumber floor that is installed as a basement floor instead of concrete.

Stucco — Refers to an outside plaster finish made with Portland cement as its base.

Stud — A vertical wood framing member, also referred to as a wall stud, attached to the horizontal sole plate below and the top plate above.

Stud Framing — A building method that distributes structural loads to each of a series of relatively lightweight studs.

Sump Pump — A submersible pump in a sump pit that pumps any excess ground water to the outside of the home.

Suspended Ceiling — A ceiling system supported by hanging it from the overhead structural framing.

Sway Brace — Metal straps or wood blocks installed diagonally on the inside of a wall from bottom to top plate, to prevent the wall from twisting, racking, or falling over domino fashion.

Switch — A device that completes or disconnects an electrical circuit.

T & G (Tongue and Groove) — A joint made by a tongue (a rib on one edge of a board) that fits into a corresponding groove in the edge of another board to make a tight flush joint.

Tab — The exposed portion of strip shingles defined by cutouts.

Tail Beam — A relatively short beam or joist supported in a wall on one end and by a header at the other.

Tax Lien — Claim against a property for unpaid taxes.

T Bar — Ribbed, "T" shaped bars with a flat metal plate at the bottom that are driven into the earth.

Teco — Metal straps that are nailed and secure the roof rafters and trusses to the top horizontal wall plate.

Tempered — Strengthened. Tempered glass will not shatter or create shards, but will pelletize like an automobile window.

Tenon — Projection at the end of the board that inserts into a mortise.

Termites — Wood eating insects that superficially resemble ants in size and general appearance, and live in colonies.

Termite Shield — A shield of galvanized metal, placed in or on a foundation wall or around pipes to prevent the passage of termites.

Terra Cotta — A ceramic material molded into masonry units.

Threshold — The bottom metal or wood plate of an exterior door frame.

Tie Beam — Also collar beam; ties together the principal rafters of a roof.

Tieback Member — A timber perpendicular to a retaining wall that ties it to a dead man buried in the ground.

Time and Materials Contract — A construction contract which specifies a price for different elements of the work such as cost per hour of labor, overhead, profit.

Tinner — Another name for the heating contractor.

Tip Up — The downspout extension that directs water (from the home's gutter system) away from the home.

Title — Evidence of a person's legal right to ownership of a property.

Toenailing — To drive a nail in at a slant. Method used to secure floor joists to the plate.

Top Chord — The upper or top member of a truss.

Top Plate — Top horizontal member of a frame wall supporting ceiling joists, rafters, or other members.

Transmitter (Garage Door) — The small, push button device that causes the garage door to open or close.

Trap — A plumbing fitting that holds water to prevent air, gas, and vermin from backing up into a fixture.

Tray Ceiling — Raised area in a ceiling that looks like a small vaulted ceiling.

Treated Lumber — A wood product which has been impregnated with chemical pesticides.

Trim (Plumbing, Heating, Electrical) The work that the mechanical contractors perform to finish their respective aspects of work, and when the home is nearing completion and occupancy.

Trim — Interior: The finish materials in a building, such as moldings applied around openings (window trim, door trim) or at the floor and ceiling of rooms (baseboard, cornice, and other moldings).

Trimmer — The vertical stud that supports a header at a door, window, or other opening.

Troweling — Process of smoothing concrete, used after floating.

Truss — An engineered and manufactured roof support member with "zig-zag" framing members.

Truth-In-Lending Act — Federal law requiring written disclosure of the terms of a mortgage by a lender to a prospective borrower within three business days of application.

Tub Trap — Curved, "U" shaped section of a bathtub drain pipe that holds a water seal to prevent sewer gasses from entering the home through tubs water drain.

Turnkey — A term used when the subcontractor provides all materials (and labor) for a job.

Turpentine — A petroleum, volatile oil used as a thinner in paints and as a solvent in varnishes.

Undercoat — A coating applied prior to the finishing or top coats of a paint job.

Underground Plumbing — The plumbing drain and waste lines that are installed beneath a basement floor.

Underlayment — A 1/4" material placed over the subfloor plywood sheeting and under finish coverings, such as vinyl flooring, to provide a smooth, even surface.

Underwriting — The process of verifying data and evaluating a loan for approval.

Union — A plumbing fitting that joins pipes end-to-end so they can be dismantled.

Usury — Interest charged in excess of the legal rate established by law.

Utility Easement — The area of the earth that has electric, gas, or telephone lines.

Valley — The "V" shaped area of a roof where two sloping roofs meet. Water drains off the roof at the valleys.

Valley Flashing — Sheet metal that lays in the "V" area of a roof valley.

Valuation — An inspection carried out for the benefit of the mortgage lender to ascertain if a property is a good security for a loan.

Valuation Fee- — The fee paid by the prospective borrower for the lender's inspection of the property.

Vapor Barrier — A building product installed on exterior walls and ceilings under the drywall and on the warm side of the insulation.

Variable Rate — An interest rate that will vary over the term of the loan.

Veneer — Extremely thin sheets of wood.

Vent — A pipe or duct which allows the flow of air and gasses to the outside.

Vermiculite — A mineral used as bulk insulation and also as aggregate in insulating and acoustical plaster and in insulating concrete floors.

Veterans Administration (VA) — A federal agency that insures mortgage loans with very liberal down payment requirements for honorably discharged veterans and their surviving spouses.

Visqueen — A 4 mil or 6 mil plastic sheeting.

Voltage — A measure of electrical potential. Most homes are wired with 110 and 220 volt lines.

Wafer Board — A manufactured wood panel made out of 1" to 2" wood chips and glue.

Waste Pipe and Vent — Plumbing plastic pipe that carries waste water to the municipal sewage system.

Water Board — Water-resistant drywall to be used in tub and shower locations. Normally green or blue colored.

Water Meter Pit (or Vault) — The box/cast iron bonnet and concrete rings that contains the water meter.

Water Table — The location of the underground water, and the vertical distance from the surface of the earth to this underground water.

Whole House Fan — A fan designed to move air through and out of a home and normally installed in the ceiling.

Wind Bracing — Metal straps or wood blocks installed diagonally on the inside of a wall from bottom to top plate, to prevent the wall from twisting, racking, or falling over domino fashion.

Window Buck — Square or rectangular box that is installed within a concrete foundation or block wall.

Window Frame — The stationary part of a window unit; window sash fits into the window frame.

Wire Nut — A plastic device used to connect bare wires together.

Wonderboard™ — A panel made out of concrete and fiberglass used as a ceramic tile backing material.

Worker's Compensation — This is a policy or endorsement covering the contractor, subcontractor, and others who will be working on the subject property.

Wrapped Drywall — Areas that get complete drywall covering, as in the doorway openings of bifold and bypass closet doors.

Wrap-Around Mortgage — Loan arrangement in which an existing loan is combined with a new loan.

Yard of Concrete — Concrete is 3'x3'x3' in volume, or 27 cubic feet.

Yoke — The location where a home's water meter is sometimes installed between two copper pipes, and located in the water meter pit in the yard.

Z-bar Flashing — Bent, galvanized metal flashing that is installed above a horizontal trim board of an exterior window, door, or brick run.

Zone — The section of a building that is served by one heating or cooling loop because it has noticeably distinct heating or cooling needs.

Zone Valve — A device placed near the heater or cooler, which controls the flow of water or steam to parts of the building; it is controlled by a zone thermostat.

Zoning Ordinances (or Zoning Regulations) — Local law establishing Building Codes and usage regulations for properties in a specified area.

INDEX